Studies in Infrastructure and Resources Management

Edited by

Thomas Bruckner
Erik Gawel
Robert Holländer
Daniela Thrän

Volume 4

The Photovoltaic Support Scheme in Germany

An Environmental Criteria Assessment
of the EEG Feed-in Tariffs

Christoph Töpfer and Erik Gawel

Studies in Infrastructure and Resources Management

Edited by

Thomas Bruckner
Erik Gawel
Robert Holländer
Daniela Thrän

Volume 4

Studien zu Infrastruktur und Ressourcenmanagement
Studies in Infrastructure and Resources Management

Herausgegeben von / Edited by
Thomas Bruckner, Erik Gawel, Robert Holländer, Daniela Thrän

Universität Leipzig
Wirtschaftswissenschaftliche Fakultät
Institut für Infrastruktur und Ressourcenmanagement
Grimmaische Str. 12
04109 Leipzig
Tel.: +49(0) 341 / 97 33 870
Fax: +49(0)341 / 97 33 879
E-Mail: umwelt@wifa.uni-leipzig.de
http://www.wifa.uni-leipzig.de/iirm/

Bibliografische Information der Deutschen Nationalbibliothek
Die Deutsche Nationalbibliothek verzeichnet diese Publikation
in der Deutschen Nationalbibliografie; detaillierte bibliografische
Daten sind im Internet über http://dnb.d-nb.de abrufbar

ISBN 978-3-8325-3485-1
ISSN 2191-9623

Logos Verlag Berlin GmbH
Comeniushof, Gubener Str. 47
10243 Berlin
Tel.: +49(0) 30 / 42 85 10 90
Fax: +49(0) 30 / 42 85 10 92
http://www.logos-verlag.de

Contents

List of Figures

List of Tables

Formulas

Abbreviations

AGEE .... German Working Group Renewable Energies - *Arbeitsgemeinschaft Erneuerbare Energien*

AHP .... Analytical Hierarchy Process

ASAY .... Average Specfic Annual Yield

AusglMechV ..... Ordinance on a Nationwide Equalization Mechanism - *Ausgleichmechanismusverordnung*

BAnz .... German Federal Gazette - *Bundesanzeiger*

BDEW .... German Federal Association of the Energy and Water Industry - *Bundesverband der Energie- und Wasserwirtschaft*

BEE .... German Federal Association of Renewable Energies - *Bundesverband Erneuerbare Energie e.V*

BIPV .... Building Integrated Photovoltaics

BMBF .... German Federal Ministry of Education and Research - *Bundesministerium für Bildung und Forschung*

BMU .... German Federal Ministry of the Environment - *Bundesminsiterium für Umwelt, Naturschutz und Reaktorsicherheit*

BMWi .... German Federal Ministry of Economics and Technology - *Bundesministerium für Wirtschaft und Technologie*

BoS .... Balance of System

BSW-Solar .... German Federal Association of the Solar Industry - *Bundesverband Solarwirtschaft*

c-Si .... Crystalline Silicon

CdTe .... Cadmium Telluride

CED .... Cumulative Energy Demand

CIGS .... $Cu(In, Ga)Se_2$, Copper(Indium,Gallium)Diselenide

MEA	Millennium Ecosystem Assessment
Mono-Si	Mono-Crystalline Silicon
Multi-Si	Multi-Crystalline Silicon
NEA	Net Emission Avoidance
NRC	United States National Research Council
NREL	United States National Renewable Energy Laboratory
PV	Photovoltaics
R&D	Research and Development
REN21	Renewable Energy Policy Network for the 21th Century
SPD	Social Democratic Party of Germany - *Sozialdemokratische Partei Deutschlands*
StrEG	Electiricy Feed-In law - *Stromeinspeisungsgesetz*
Tab.	Table
TGC	Tradable Green Certificate
TSO	Transmission System Operator
UBA	German Federal Environmental Agency - *Umweltbundesamt*
UCTE	Union for the Co-ordination of Transmission of Electricity
UNEP	United Nations Environment Programme
UNFCCC	United Nations Framework Conference on Climate Change
VDI	Association of German Engineers - *Verein Deutscher Ingenieure*
WBCSD	World Business Council for Sustainable Development
WBGU	Academic Advisory Council of the German Government on Global Environmental Change - *Wissenschaftlicher Beirat der Bundesregierung Globale Umweltveränderungen*

Symbols and Units

R_{prim} .... Conversion Factor for Primary to End Energy (MJ_{prim}/kWh_{end})

Rem_{act} .... Actual Remuneration $(\text{€}/kWp)$

Rem_{ref} .... Reference Remuneration $(\text{€}/kWp)$

SR .... Space Requirement (m^2/kWp)

T .... Period of time (years)

V_{best} .... Best Value in the Normalization Matrix

$V_{i,j}$ .... Current Value of Installation Option i and Technology j

V_{worst} .... Worst Value in the Normalization Matrix

r .... Interest rate (%)

t .... Point in time (year)

ΔRem .... Difference in Remuneration $(\text{€}/kWp)$

ΔRem_{env} .... Difference in Remuneration with an Environmental Bonus $(\text{€}/kWh)$

CO_2eq .... Carbon Dioxide Equivalent

GWp .... Gigawatt Peak

kWh_{end} .... Kilowatt Hours End Energy

kWp .... Kilowatt Peak

MJ_{end} .... Megajoule End Energy

MJ_{prim} .... Megajoule Primary Energy

MWp .... Megawatt Peak

Wp .... Watt Peak

1 Introduction

1.1 Approaching the Issue

Climate change and its associated impacts are seen as a current and future threat to society, human health and eco-systems from a local to global scale (cf. EEA 2012, p. 15). To limit its negative consequences governments commonly agreed within the United Nations Framework Conventions on Climate Change (UNFCCC) to constrain the rise of global mean temperature to 2°C (cf. UNFCCC 2011, p. 3). In order to comply with this goal in a cost efficient way greenhouse gas (GHG) emissions have to be reduced substantially and the reduction should start the latest in the mid of the current decade (cf. IEA 2010, p. 46; WBGU 2009, pp. 7, 43). The energy sector represents a key determinant for mitigating climate change because 84% of global GHG emissions are energy related. From this amount 41% are attributable to the generation of electricity (cf. IEA 2010, pp. 54, 102).

There are in principle three options for a decarbonization of the electricity generation system, being either the ongoing deployment of carbon dioxide (CO_2) intensive energy carriers in combination with carbon capture and storage technologies, the use of nuclear power or the expansion of renewable electricity technologies (cf. WBGU 2009, p. 43). Energy efficiency improvements are considered essential but no single solution for limiting climate change (cf. Bruckner et al. 2010, p. 193; WBGU 2009, p. 45). Scenarios of integrated assessment models, which describe a development of the energy system in order to achieve a dynamic cost-efficient[1] fulfillment of climate targets, reveal that a portfolio of renewable technologies is a necessity for reaching this long-term goal (cf. e. g. IPCC 2012, Chapter 10; IEA 2010, p. 47; 2008, p. 27). In addition, renewable sources for electricity generation in particular, like direct solar or wind among others, show an abundant availability (or technical potential[2]) being able to cover current and projected electricity demand (cf. IPCC 2012, pp. 183-184).

Besides climate change mitigation, there is a multitude of other very important but often neglected positive effects associated with the deployment of renewable electricity technologies as a substitution for those based on fossil fuels like hard coal, lignite, oil, gas or on uranium. Among these are avoided air pollution, crop losses, material damage, biodiversity loss, environmental damages associated with fossil resource extraction (e. g. mining, oil production), accidents and disasters (e. g. oil spill, nuclear material release from power plants), waste storage problems (e. g. nuclear waste) as well decreased import dependency, increased employment opportunities and access to electricity (cf. IPCC 2012, pp. 878-880; BMU 2012c, p. 51; Lehmann and Gawel 2013; Madlener and Stagl 2005).

Based on this, the European Union (EU) and its member states set ambitious targets for the reduction of GHG emission and the increased deployment of renewable electricity technologies. The "2020 package" of the EU for instance aims to decrease GHG emissions by 20% (compared to 1990) and increase the share of renewables in energy consumption to 20% by 2020 (cf. European Commission 2008). To reach them the directives 2003/87/EC, implementing the European emission trade scheme (EU ETS), 2009/28/EC and its predecessor 2001/77/EC, for the promotion of renewable energies, were enacted. The two latter directives in particular call for support policies comprising renewable quotas, feed-

---

[1] Dynamic and static cost-efficiency have to be separated. The former refers to an objective fulfillment at least cost over time incorporating learning effects and inter-temporal dependencies. The latter in turn means an objective achievement at least cost with a given technology. (cf. EFI 2013, p. 49)

[2] The technical potential is defined as "the amount of renewable energy output obtainable by full implementation of demonstrated technologies or practices. No explicit reference to costs, barriers or policies is made." (IPCC 2012, p. 963)

in tariffs (FIT), premium payments, investment subsidies or tax based support among others. In Germany the renewable energy sources act (*Erneuerbare Energien Gesetz* - EEG) is the main policy tool for promoting renewable electricity generation by setting up a fixed FIT scheme. It targets to establish a sustainable energy supply with regards to climate and environmental protection, limited energy supply cost, fossil energy resource savings and the promotion of renewable technologies (§ 1 (1) EEG 2012-PV) by increasing the share of renewables in successive steps to 80% in 2050.

Currently renewables contribute with 20.3% to the German electricity mix (date 2011) (cf. BMU 2012a). Photovoltaics (PV) in particular showed an enormous growth. The installed PV capacity was enlarged from the year 2000 until 2011 by approximately 24.8 GWp, from which 14.9 GWp were installed between 2010 and 2011 (cf. REN21 2012, p. 101). With the EEG Germany became the largest PV market with a fraction of 35.6% of the worldwide installed PV capacity (date 2011) (cf. REN21 2012, p. 48).

However, the rapid expansion also comes at a cost. The recent significant increase in the EEG reallocation charge borne by the electricity consumers on top of every consumed kWh triggered a political debate about limiting the costs of the German energy system transition (*Energiewende*). Especially the support for PV is challenged because of the comparably high FITs granted to PV system operators (cf. BMWi and BMU 2012a, 2012b, p. 31). Consequently, economic efficiency assessment and the determination of the "right" level of support have been a major subject within the EEG monitoring procedure and reason for law amendments (cf. Bundesregierung 2011; BMU 2007; Bundesregierung 2002a). On the contrary, it is interesting to observe that other efficiency criteria have received rather limited attention. The environmental goals of the EEG are, in theory, highly prioritized by the legislator (cf. EEG-Explanation 2004, p. 12) but the assessment of their efficient achievement is not explicitly addressed (cf. Reichmuth 2011, pp. 231-293; Staiß et al. 2007, pp. 339-347). Madlener and Stagl (2005) state in this sense that *"when designing policy instruments for more sustainable energy futures [...] the aim from a sustainability point of view ought to be to generate the lowest possible adverse socio-ecological economic impact (SEE) per unit of electricity generated, while ensuring a certain degree of economic efficiency."* (p. 149) Based on this, the present study aims to establish a connection between the economic investment stimulus of renewable support schemes and the environmental performance of their supported technologies taking the EEG PV FIT scheme as an example.

Therefore, the research question addressed in this study is formulated as follows: *Are environmental criteria reflected in the German PV support scheme?* It is approached through three workable sub-research questions on which the upcoming chapters are based:

a)   Which exemplary criteria and indicators could be used to assess the EEG's environmental efficiency?

b)   Is the FIT system applied by the EEG providing investment incentives that are consistent with these environmental criteria?

c)   Which potential improvements can be derived from the analysis and what are their implications for the EEG and PV support schemes in general?

Since economic efficiency is already a major topic of investigations, it is explicitly not the aim of this study to assess which level of support for PV electricity generation might be appropriate. It is rather scrutinized if the EEG provides incentives for the environmental efficient installation of PV capacity, how PV support schemes can achieve lower environmental impacts from the same operating capacity and what the economic impacts of such an environmental orientation could be.

1.2 Methodology

Based on the research questions stated above this study aims to provide a qualitative, comparative analysis of the investment incentives induced by the EEG PV FIT scheme and the corresponding environmental impacts for a set of PV reference plants. Comparisons between different kinds of renewables or between PV and conventional electricity generation are out of scope in this study. For the envisaged analysis the term environmental efficiency is defined. It indicates that a renewable support scheme, providing investment incentives compatible with the environmental performance of the promoted technologies, can achieve the same amount of renewable electricity generation or installed capacity at an absolutely lower level of environmental impacts, or, *vice versa*, lead to an increased electricity output/installed capacity at an unchanged level environmental impacts.

In order to determine the environmental impacts, criteria and indicators for their measurement are derived from an in-depth analysis of the EEG, its objectives and support rationales. Life cycle assessment (LCA) according to the European norm 14040:2006 of the International Organization for Standardization (EN ISO 14040:2006) and EN ISO 14044:2006 is used as a methodological framework for estimating the environmental impact of eight exemplary PV installation options combined with three currently market dominating module technologies. Included in the assessment are five roof-mounted, two facade-mounted and one open ground PV installation option each equipped with either multi- (Multi-Si) and mono-crystalline silicon (Mono-Si) or Copper(Indium,Gallium)Diselenide (CIGS) thin-film module technologies, resulting in 24 considered installation/technology combinations. It is aimed to reveal environmental impacts concerning climate change (carbon footprint), lifetime GHG emission avoidance in relation to direct space occupation (net mission avoidance - $NEA_{space,LT}$) and primary resource consumption (Energy Pay-back Time - EPBT). The life cycle impact assessment (LCIA) is based on a life cycle inventory (LCI) model previously elaborated in Töpfer (2012) comprising on-site collected producer data of Hanwha Q CELLS and Solibro respectively for the years 2011 and 2010.

Counting with the environmental performance as one comparison component, the investment incentives induced by the EEG PV FIT scheme are derived from the FIT rates of the EEG legislation in 2000, 2004, 2009, 2012 and 2013 as the second component. This is done for a set of reference PV plants corresponding to those analyzed in the LCA and covering all size classes differentiated by the EEG. Based on the methodology for stipulating the PV FIT level outlined by Schmidt (2012) and the German Federal Ministry of the Environment (BMU 2007, p. 172), it is assumed that the observed EEG FITs of the specific PV plants under consideration are cost covering and provide an appropriate rate of return for the plant operator or investor. In order to be comparable to each other, both, the environmental indicator results and specific investment incentives are normalized to a scale from 0 to 100. Out of this, it is investigated whether or not the EEG reflects environmental criteria in its PV remuneration scheme.

Moreover, literature research is used to underpin the findings of the LCA study in a sensitivity analysis and a comparative literature review as well as to classify the EEG PV FIT scheme within the broader context of renewable promotion policies. Out of the research on support policies for renewable electricity generation possible implications from the carried out comparison are critically scrutinized.

1.3 Structure of the Study

The present study is structured in six chapters that are oriented on the above-mentioned successive sub-research questions. Chapter 2 sets up the theoretical framework of this study by introducing policy options for renewable electricity promotion, rationales to implement them and inherent design possibilities (Section 2.1). Moreover, the EEG PV FIT scheme as the object of analysis is presented in Chapter 2. A brief examination of the development until its current state and the progress of its goals (Section 2.2), the reasoning behind its FIT level determination (Section 2.3) as well as the EEG induced renewable electricity capacity enlargements, corresponding effects on GHG avoidance and accruing support cost (Section 2.4) are included. The most important findings are then summarized in Section 2.5, where important shortcomings in the current environmental assessment and FIT level determination of the EEG are revealed, justifying the further research. In Chapter 3 environmental indicators for the intended comparison to the EEG PV FIT scheme are developed. Therefore, Section 3.1 explains the selection process of criteria and corresponding indicators for the subsequent environmental LCA. Section 3.2 gives insight in the previously developed LCA model (Töpfer 2012). Afterwards the environmental indicator results are presented (Section 3.3), checked for robustness in a sensitivity analysis (Section 3.4) and compared to the findings of other authors in a comparative literature review (Section 3.5). Chapter 4 aims to combine the EEG PV FIT scheme and the revealed environmental indicators results. To achieve this, a normalization of the obtained results is applied (Section 4.1) and a base for comparison developed (Section 4.2). The actual comparison is presented Section 4.3 and Section 4.4. The discussion in Chapter 5 then critically reflects on the previously obtained results and investigates possible effects that the analysis could have on the renewable support policy design (Section 5.2), comprising the consideration of environmental criteria in the prevailing EEG FIT scheme, other support policies and conceivable side effects that might occur in combination with other energy policy instruments. Finally, Chapter 6 summarizes the main conclusion of this study with regards to the raised research question and outlines possibilities for future research and methodological alterations.

2 Supporting Photovoltaics - The Case of the German EEG

The present chapter aims at introducing renewable electricity support schemes and possible policy design options in general (Section 2.1) and subsequently assesses the German EEG based on the theoretical findings (Section 2.2). Based on the in-depth EEG PV support analysis the rationale behind its FIT determination is assessed in Section 2.3 and its induced effects and costs are outlined (Section 2.4). The chapter concludes by summarizing and evaluating the most important findings for the subsequent analysis in Section 2.5.

2.1 General Promotion Possibilities of Renewable Electricity Generation

This section draws on general promotion possibilities of renewable electricity generation. It starts with an investigation of reasons for the implementation of support policies (Section 2.1.1). Afterwards a classification of policy options is presented (Section 2.1.2) and the most prominent support schemes of quota obligation and FITs are explained in more detail (Section 2.1.3 and Section 2.1.4).

2.1.1 Rationale of Promoting Renewables

Energy provision from renewable sources is seen as one of the main pathways to comply with the commonly agreed upon 2°C goal of maximal global mean temperature rise (cf. IPCC 2012, pp. 169, 177) and thereby to constrain the negative effects associated with climate change (cf. IPCC 2007, pp. 48-54; IEA 2008, p. 27). Therefore, and among other reasons like improved environmental effects (water consumption, air pollution, land use, biodiversity etc.), increased energy security, increased access to energy, improved health effects or employment and economic growth projections, enlarging the share of renewable energies in the provision of both, primary and final energies, is considered socially beneficial and should be striven for (cf. IPCC 2012, pp. 878-880; Klein et al. 2008, p. 5; IEA 2008, p. 27). In the electricity sector an increased provision from renewables with low variable costs could in addition reduce the electricity prices for consumers in the long-term because of merit-order effects that occur in combination with priority grid feed-in of renewable electricity (cf. Sensfuß 2011a, pp. 3-5).[1]

The Intergovernmental Panel on Climate Change (IPCC 2012, pp. 194, 872) justifies governmental intervention by the presence of market failures which would produce an unfavorable low amount of renewable energy deployment and CO_2 reduction under (imperfect) market conditions. Two market failures are emphasized in particular: firstly, the investor's underestimation of the future benefits resulting from technological learning and spill-over effects generated by the new technology in combination with possible high upfront investments and, secondly, the uninternalized external social costs accruing from GHG emissions. The first argument calls for governmental support to increase technological market diffusion via renewable energy policies while the latter targets on an internalization by setting a price on GHG emission, e. g. by the EU ETS or an environmental tax (Pigouvian Tax) (cf. IPCC 2012 p. 872). Thus, each market failure should in theory be tackled by one policy to assure an optimal social outcome (cf. IPCC 2012, p. 916), i. e. avoiding welfare losses. Lehmann and Gawel (2013) argue that if the only policy aim were to avoid climate change and only the climate externality existed, pricing GHG emissions would suffice as a single policy to

---

[1] The impact of the merit-order effect is diversely discussed in literature (cf. e. g. Mennel 2012) and is mentioned here, like other aspects, as a possible justification of renewable energy support strategies.

address the issue of climate change most efficiently, implying that additional renewable promotion strategies become unnecessary.

Generally, in case of multiple market failures (e. g. spill-over effects, asymmetric information, environmental damages apart from climate change) or high induced transaction costs (e. g. by heterogeneous marginal pollution damages or abatement costs, non-compliance of polluters) a policy mix might increase efficiency and should be preferred over a single pollution policy (cf. Lehmann 2012). Furthermore, there is a multitude of pursued objectives connected to the promotion of renewables providing a political justification for policy implementation. Besides the market failures established by learning and spill-over effects (cf. IPCC 2012, pp. 872), Lehmann and Gawel (2013), discussing the interaction of the EU ETS and renewable promotion policies, identify occurring policy failures that can justify the promotion of renewables also in economic terms. They argue that governments might not succeed in overcoming prevailing market failures so that a renewable support schemes can provide a second-best policy option or that policies themselves induce further market distortions (e. g. provide subsidies to non-renewable electricity generation or create investment uncertainties). Concerning the first argument, the incomplete internalization ability of external costs from GHG emissions by the EU ETS, being theoretically the best-policy option, is pointed out. Because of highly uncertain damage costs of GHG emissions, biased certificate allocation procedures and the negligence of other detrimental environmental effects in the EU ETS its efficient climate change mitigation could be challenged (cf. Lehmann and Gawel 2013). Moreover, it is argued that emission caps in an ETS might be fixed in combination with anticipated GHG reduction from the promotion of renewables, thus indirectly decreasing the amount of emissions (cf. Lehmann and Gawel 2013; IPCC 2012, p. 917). Another argument in favor of a policy mix for mitigating climate change are path dependencies in carbon intensive electricity generation structures, caused by inertia in technology adaption, long investment cycles in combination with high investment amounts, low product differentiation possibilities and inert institutional change in the electricity sector (cf. Lehmann and Gawel 2013). Hence, a policy fostering technological diversity and the overcoming of current market barriers for renewable electricity technologies can also help mitigating climate change because it enables to make future use of technologies with high mitigation potential (cf. IPCC 2012, p. 914; Philibert 2011, p. 10).

In general, policies supporting renewables should aim to create a harmonized "enabling environment" to facilitate change and innovation in the energy system and overcome prevailing barriers for the deployment of renewable technologies, e. g. by providing investment security (cf. IPCC 2012, pp. 870, 917-919), additionally setting comprehensive long-term objectives and being able to provide sufficient flexibility to learn from experience (cf. IPCC 2012, p. 870).

Especially technological learning is seen as a key determinant for the application and introduction of renewable energy support (cf. Lehmann and Gawel 2013; IPCC 2012, p. 932; del Río 2012). Therefore this concept is outlined here in more detail. It is also an important driver of the EEG's FIT policy (see below). Learning rates reflect the decrease in cost with a doubling of the cumulative installed capacity (cf. Kersten et al. 2011). According to IPCC (2012 p. 846) learning effects can be triggered by research and development (R&D) activities, improvements in the production process, user experience, interactions and spill-overs, upsizing of technologies and economies of scale in mass production. In case of PV systems, learning rates are estimated by the European Photovoltaic Industry Association (EPIA 2011) to approximately 20%, which is also verified by IPCC (2012 pp. 380, 848). EPIA (2011 p. 70) predicts the technological learning of PV to decrease in the future to 18% from 2020, 16% from 2030 and 14% from 2040-2050 on. Technological learning could thereby justify the promotion of market penetration in a way, that once those technologies become competitive, the learning investments pay off due to lower generation costs compared to the conventional technology (cf. IPCC 2012, p. 849; EPIA 2011, p. 34). However, because learning effects can be found in different parts of the value chain, the applicability of support instruments should be tailored according to the state of technological development and interdependencies with other promotion strategies and, thus, include "supply push" as well as "demand pull" policies (cf. IPCC 2012, pp. 197, 851; IEA, 2010 p. 50). For an infant technology like e. g. fuel cells, R&D financing might be appropriate, whereas for a technology like PV, which is already beyond the demonstration stage but exhibits a high cost gap to conventional electricity production and is therefore "trapped" in niche markets, a market diffusion support induced by economic incentives can be more effective (cf. IEA 2010, p. 7). The IPCC (2012 p. 932) summarizes that, in order to ensure effectiveness, R&D support (supply push) should be complemented with policies ensuring the commercialization of technologies (demand pull) in order to create a market for them. This market could in turn offer positive feedback to investments in R&D and learning effects.

By now 118 countries around the world have set targets for renewable energy deployment, from which 109 count with policies towards renewable electricity generation (cf. REN21 2012, pp. 65-66). The German government for instance aims at increasing the share of renewables in gross electricity consumption by 2050 to 80% (BMU 2011, p. 5). On a European scale the directive 2009/28/EC on renewable energies demands 20% of gross end energy consumption to be supplied by renewable sources by 2020.

2.1.2 Policy Options

In order to achieve these ambitious aims for renewable energy deployment, governments are equipped with a set of policy options which are distinguishable on a broad scale in fiscal incentives, public finance and regulation (cf. IPCC 2012, p. 197). However, since this study focuses on the power sector in particular a more specified strategy classification of Haas et al. (2011) is presented (see Tab. 2.1). At first, the authors differentiate direct and indirect instruments, where the former are tools for short-term stimulation of renewable electricity technology promotion and the latter are measures that indirectly facilitate the proliferation of renewables, e. g. by setting a tax on the use of non-renewable

Tab. 2.1: Classification of Renewable Energy Promotion Strategies; Source: Haas et al. 2011

		Direct		Indirect
		Price-driven	**Quantity-driven**	
Regulatory	Investment focussed	– Investment subsidies – Tax credits – Low interest/soft loans	– Tendering system for investment grants	– Environmental taxes – Simplification of authorization procedures – Connexion charges – Balancing costs
	Generation based	– (fixed) feed-in tariffs – Fixed premium system	– Tendering system – Quota obligations based on tradable green certifictates	
Voluntary	Investment focussed	– Shareholder programmes – Contribution programmes		– Voluntary agreements
	Generation based	– Green tariffs		

energy sources, pricing GHG emissions or removing subsidies for non-renewable power generators as well as tax exemptions. Within the direct promotion strategies price driven and quantity driven approaches, supporting either investments in capacity or are subsidies for renewable electricity generation, can be identified (cf. Jenner et al. 2013; Haas et al. 2011; IEA 2008, p. 92). A further distinction is made in regulatory and voluntary instruments, with the latter depending *"on consumers' willingness to pay premium rates for green electricity"* (Haas et al. 2011, p. 1011). Among the regulatory investment focused strategies are investment subsidies, tax credits, low interest loans and tendering or bidding systems for investment grants. Generation based regulatory approaches comprise different FIT systems as price driven strategies and tendering or quota systems as quantity driven strategies.

Additionally, Boute (2012) discusses a novel option for promoting renewable electricity generation via specially designed capacity markets, i. e. a remuneration of the installed capacity and its availability instead of the amount of electricity generated or fed into the grid. Since FIT schemes and quota obligations are most popularly used by governments to accelerate the market diffusion and thus enlarging the share of renewables in the power sector (cf. REN21 2012, p. 66; IPCC 2012, p. 874; Haas et al. 2011; Couture and Gagnon 2010; Fouquet and Johansson 2008; Ragwitz et al. 2007), their different design options will be briefly explained in the following, with the purpose of recognizing specific aspects in the FIT support scheme constituted by the German EEG (Section 2.2) and building a theoretical framework for the upcoming discussions (Chapter 5).

However, in literature there is a broad debate about what is the most dynamically efficient/effective support scheme that can be deployed (cf. e. g. Becker and Fischer 2013; Boute 2012; Haas et al. 2011; Klessmann 2008; Fouquet and Johansson 2008; Ragwitz et al. 2007), which design parameters and characteristics play a crucial role (cf. e. g. Jenner et al. 2013; Kim and Lee 2012; del Río 2012; Couture and Gagnon 2010) and how different promotion strategies and their peculiarities can counteract with other policy instruments, like cap and trade schemes, in order to provide GHG emission abatement most cost-efficiently (cf. e. g. Lehmann and Gawel 2013; Lehmann 2012; Philibert 2011; Frondel et al. 2010). This chapter only presents basic options for possible renewable electricity promotion policy designs and will leave some aspects of the mentioned debate for the discussion where additionally the findings of this study are integrated (see Section 5.2).

2.1.3 Quota Obligations

Quota obligations, as a quantity driven instrument, politically fix a desired level of renewable electricity that has to be provided by power generators. The corresponding price is then defined by the market (cf. Lipp 2007). The necessary provision of installed capacity or renewable electricity can be allocated either through tendering processes where capacities are auctioned via competitive bidding or tradable green certificate (TGC) systems in which a market for "green" electricity is created. In the former case the winner of the bid can receive favorable investment conditions per kWp installed capacity or a guaranteed tariff per delivered kWh fixed over a certain time.[2] (cf. Haas et al. 2011)

Within a TGC system an obliged group of the electricity supply chain (e. g. retailers, generators or consumers) have to present an amount of certificates according to the goal fixed by the legal authority (cf. Haas et al. 2011). *"The purpose of the TGC systems is to reward the greenness of renewable electricity by the creation of a liquid market in certificate paper in addition to sales of physical power"* (Verbruggen and Lauber 2012, p. 638). Certificates can either be bought on the market or earned by renewable electricity generation and subsequently sold on the market, generating an additional revenue over the market price for electricity (cf. Haas et al. 2011). Thereby the certificate price is determined by the rules of supply and demand, signifying that with reaching the fixed expansion target, prices would decrease to zero (cf. Fouquet and Johansson 2008). Generally, penalties are introduced if the mandatory capacity expansion quotas are not met by the renewable electricity producers (cf. Verbruggen and Lauber 2012). Applying a TGC scheme provides flexibility in meeting the quota, because renewable electricity can be self-generated, imported from somewhere else or fulfilled by buying certificates on the market (cf. Mendonca et al. 2010, p. 151). In theory support costs are kept low because of market oriented price determination and distribution of certificate prices among electric utilities and electricity consumers (cf. Mendonca et al. 2010, p. 152).

Verbruggen and Lauber (2012) argue that the flexible revenues, depending on the amount of certificates in the market, can deteriorate the operability of existing plants or the abandonment of new projects and consequently hinder the renewable capacity expansion. Haas et al. (2011) describe the TGC strategy *"all in one basket promotion"*, deploying generally the most cost-efficient renewable electricity options, or those close to maturity, which could in turn prevent currently expensive technologies (like PV) from developing and learning (cf. Fouquet and Johansson 2008). According to Haas et al. (2011) a solution to that problem could be the split up of certificate markets to facilitate a technology specific pro-

---

[2] This system is discussed in more detail in Section 5.2

motion (technology banding) or a weighting of certificates according to a technological preference relation. However, the first option would reduce the market's liquidity and the second one would entail the problem of justifying the chosen weights. Moreover there might be the chance of arising monopolies of large generators in the electricity market and subsequent price influencing possibilities (cf. Fouquet and Johansson 2008). This can be partly explained because of prevailing market risks of the conventional electricity and certificate market, with which experienced electricity market actors can cope more easily than new entrants, as well as the possibility of buyouts of unprofitable renewable electricity plants by larger companies due to the insecure investment environment (cf. Verbruggen and Lauber 2012).

2.1.4 Feed-in Tariffs

On the other hand, FIT schemes, classified as a regulatory, generation based and price-driven promotion strategy (Tab. 2.1), face generators of renewable electricity with a usually technology specific fixed price over a determined period of time, while the costs that accrue are normally transferred to the end consumer of electricity (cf. Couture and Gagnon 2010). Consequently, FIT schemes promote the competitiveness of specific renewable technologies relative to conventional sources of electricity generation and thereby enhance their attractiveness to multiple potential investors (cf. Jenner et al. 2013; Couture and Gagnon 2010). They are the most widely used policy tool to promote renewable electricity generation (cf. Couture and Gagnon 2010; del Río 2012), with at least 65 countries having adopted a FIT policy to promote renewables (cf. REN21 2012, p. 66). There is a variety of conceivable design options for FIT schemes that will be summarized in the following paragraphs. Before going into detail, it should be noticed that the exact definition of what a FIT is and what distinctions to other incentive driven support schemes are, are often vague or unclear (cf. Becker and Fischer 2013). Becker and Fischer (2013) argue that in general tendering systems, offering auction based tariffs, are not considered as FITs (compare to Tab. 2.1) despite the fact that, once the quantity is allocated, both support mechanisms are quite similar. In this chapter the classification possibilities of different FIT design options will be an assembly of the remarks of Mendonca et al. (2010), Klein et al. (2008), Ragwitz et al. (2007), Couture and Gagnon (2010) and del Río (2012) seeing FITs as a purely price-driven policy.

Couture and Gagnon (2010) propose a main differentiation between FIT schemes according to their market dependency. The market dependent version is generally referred to as premium price feed-in, because it guarantees a premium payment in addition to the market price (cf. IEA 2008 p. 93). The total premium price received by the plant owner consequently fluctuates with the electricity market price prevalent when feeding the renewable electricity into the grid, entailing a competition among electricity producers and generating an incentive to feed-in electricity in times of peak demand, i. e. high spot-market prices (cf. Couture and Gagnon 2010; IEA 2008, pp. 92-93). Within this design option purchase obligations are generally not designated. Additionally, the fixed premium option can bear the risk that remunerations for the plant owners are either too low, i. e. impairing the amortization of investment cost, or too high, i. e. entailing high social cost for the deployment of renewables. To remedy these shortcomings a cap and/or a floor level for remunerations can be added so that investment security is enhanced. Moreover, there is the possibility to set the premium received relative to the market price. Premium based tariffs with or without caps and floors are currently used in countries like Spain, Czech Republic, Estonia and Denmark. (cf. Couture and Gagnon 2010)

In contrast, market independent FITs guarantee a total price as a remuneration for renewable electricity generation over a contracted period of time, with the tariff level usually being based on the costs of generation from a specific technology. Within this option purchase obligations, that guarantee the complete remuneration of renewable electricity produced, are generally assured. There is the possibility of adapting this fixed price model to account for inflation (full or partially) over the contract time by accordingly adapting the fixed FIT rates. By not doing so, plant owners are confronted on the one hand with steadily decreasing returns in real terms but can on the other hand plan their investment with more security as all future cash-flows are known. Germany, for instance, deploys the fixed price model without direct inflation adjustment and Spain adjusts the fixed FIT completely to the inflation rate. Furthermore, the basic fixed FIT model can be varied by paying a higher tariff in the beginning of the contract which is, after a certain time or quantity of fed-in electricity, reduced to a lower rate. Thereby investors receive higher remuneration in the beginning to pay-off their capital outlay and later on, when in theory the project is close to amortization, the social payment burden is reduced. This special case of a stepped tariff (see below) can additionally contribute to cost-efficiency and setting of an adequate, non-discriminating tariff level since it avoids an overcompensation of renewable electricity facilities with good resource availability and high feed-in amounts, while still ensuring sufficient remuneration for facilities with lower yields.[3] (cf. Couture and Gagnon 2010)

A last option identified by Couture and Gagnon (2010, p. 959) is the "spot-market gap model". It is a mixture between market-dependent and independent FIT design options because it offers a fixed FIT rate for the plant owner consisting of spot-market electricity price and FIT premium. The premium is a variable component and compensates the generator with the difference between fixed FIT level and prevalent spot-market price and can be either passed on to electricity consumers or government financed. Since the produced electricity has to be sold by the generators on the spot-market this FIT model could lead to investment reluctance of smaller actors due to increased transaction cost and market barriers. (cf. Couture and Gagnon 2010)

Beyond the FIT variants compiled and discussed by Couture and Gagnon (2010); del Río (2012), Mendonca et al. (2010), Klein et al. (2008) and Ragwitz et al. (2007) specify important aspects to be considered when designing a FIT system with regard to effectiveness and economic efficiency. These are, among others: the determination of eligible plants

---

[3] For instance, a PV facility in Munich receives higher irradiation than a comparable facility in Berlin but both investors would obtain the same FIT.

for FIT remuneration, defining the FIT level as such,[4] the duration of support, the tariff revision procedure, stepped tariffs, additional premiums, tariff degression, demand orientation and financing mechanisms as well as possible net metering, forecast obligations or local promotion. The assessment of the EEG below (see Section 2.2) will outline the practical configuration of a multitude of these design parameters. For the purpose of this section the general idea behind a selection of them and the consequences of their application will be explained:

One important determinant for setting an appropriate FIT level that satisfies demands for investment security, project cost recovery and reasonable profits is the time over which the generator receives the payments. Del Río (2012) states that long durations from 15 to 20 years facilitate investment security and support currently expensive technologies with longer amortization times. On the other hand he argues that long contracts can raise consumer cost and restrict policy flexibility.

FIT revision procedures are necessary in order to check the applicability and retained economic efficiency of FITs over time (cf. Klein et al. 2008, p. 21). The tariff revision can be carried out after a certain period of time, capacity expansion volume reached (cf. Klein et al. 2008, p. 21) or for reasons of cost containment (cf. del Río 2012). Therefore most FIT schemes are already equipped with an envisaged revision procedure to ensure the system's flexibility (cf. Klein et al. 2008, pp. 21-23).

A stepped tariff can, in addition to what has been already stated by Couture and Gagnon (2010), account for differences in costs occurring within the same technology (e. g. economies of scale) by differentiating the remuneration for different plant sizes (cf. Mendonca et al. 2010, pp. 26-27; Klein et al. 2008, pp. 24-27). This is an important option used for promoting PV in the German EEG and will be described in more detail below (see Section 2.2). Furthermore a tariff discrimination can also be introduced between different day-times or seasons within a year based on the observed electricity load profile (demand orientation), giving the

---

[4] This aspect is not further mentioned here, but captured in detail in the analysis of the German EEG (see Section 2.3).

generator an incentive to feed-in at times of high electricity demand (cf. Klein et al. 2008, pp. 59-62). With this option the market orientation of the FIT scheme could be improved, but unfortunately electricity production from renewables like wind and PV is unpredictable and cannot be influenced by the plant owners (cf. Klein et al. 2008, p. 62).

There is the further option of granting a supplementary premium on top of the FIT in order to promote the fulfillment of certain criteria like e. g. high efficiency or repowering (cf. Mendonca et al. 2010, p. 53). An example for a utilization of this design parameter is the higher FIT for PV installations on buildings granted by the EEG (see below) or a higher remuneration for building integrated PV facilities in France (cf. Klein et al. 2008, p. 57). This option is highly interesting for the purpose of this study since additional environmental criteria are being developed that could be reflected by this design option (discussed in Section 5.2).

It is also possible to account for technological learning in the FIT scheme via degression rates or tariff revisions (cf. Mendonca et al. 2010, p. 49; Klein et al. 2008, p. 39). According to Klein et al. (2008, p. 40) taking learning rates into account is especially suitable for technologies not relying on fuel costs such as PV. In case of degression rates, the FIT level is decreased by a certain percentage depending on the year of power plant deployment, but once the contract is signed, the tariff level is guaranteed over the complete contract duration (cf. Ragwitz et al. 2007, p. 118). In this way, the financial burden consumers have to bear can be reduced and R&D incentives facilitated (cf. del Río 2012). Contrarily, degression rates can also provide strong incentives for rapid investments with the rationale of securing a higher FIT and thereby cause a high present social burden (cf. del Río 2012). Another disadvantage is the possible necessity for adaptions if learning rates are developing faster or slower than degression rates (cf. del Río 2012). A solution to the latter problem can be flexible tariff degressions by linking the degression rate to the market growth of a specific technology (cf. Mendonca et al. 2010, pp. 50-51). In this case degression rates are increased from their normal value if a certain percentage of market growth or amount of installed capacity is reached (cf. Mendonca et al. 2010, p. 50).

The last FIT design parameter mentioned here addresses the question of who bears the cost of the support program, i. e. describes the FIT financing mechanism. In general the cost burden is shifted to and spread amongst the consumers of electricity, not including direct financial support from governments (cf. del Río 2012; Mendonca et al. 2010, pp. 28-29; Klein et al. 2008, p. 64). The result is a higher electricity price paid by consumers, opening a discussion on how certain consumer groups are affected by this increase. Therefore, in some FIT schemes exemptions from the *"equal burden sharing"* (Klein et al. 2008, p. 10) exist for instance for energy intensive industries which see their international competitiveness at risk (cf. Klein et al. 2008, p. 65). In market dependent FIT schemes, like the spot-market gap model described above, differences from the spot-market price to the assured FIT payment can be also subsidized by governments, i. e. the taxpayer (cf. Couture and Gagnon 2010).

In summary, the most important design parameters for FITs would be (cf. del Río 2012; Mendonca et al. 2010; Klein et al. 2008):

- Premium vs. fixed tariff   - Tariff degression rate and flexibility
- Technology specificity   - Purchase obligation and priority grid access
- Level of support   - Additional premium possibilities
- Time of support   - Inflation indexing
- Revision procedures   - Demand orientation
- Financing mechanism   - Forecast obligation
- Stepped tariffs

2.2 Development and Vertices of the EEG's PV Support

The EEG is the outcome of a much longer history of renewable energy promotion in Germany. Obviously, there is no German *Sonderweg* in this regard, since on the EU level there have been established binding targets for the share of RES in the member states (20% by 2020). Furthermore, the national support schemes used in the member states tend to converge over time and the German EEG might be seen as a pioneer model in this field. As Gawel et al. (2012a) show, the idea that Germany is somehow on its own in Europe, on the fringe of the continent's mainstream on the issue, is a myth and a misconception: eleven out of twenty-seven European Union member states currently do not rely on nuclear power and most of them have committed themselves to not doing so in the future; five other European countries are joining Germany in phasing out nuclear power, among them Italy, another G8 member state. Furthermore, eighteen EU countries have implemented renewable support policies similar to those in Germany. Therefore, it is misleading to describe Germany's energy policy as a *Sonderweg* (Gawel et al. 2012a). What makes the German situation special is not the goal of a sustainable energy system as such but the context of an accelerated transition in a thoroughly industrialized economy.

This section aims to briefly outline the different predecessors and measures that led to the implementation of the EEG law in the year 2000 with a focus on support for PV. Moreover, the EEG's PV support will be analyzed over the different amendments the law has undergone. Thereby the different renewable promotion strategies and FIT design elements described in the preceding section will be recognized.

2.2.1 Development towards the EEG

The EEG was not the first law in Germany which aimed at a generous support for renewable deployment. Already in 1990 the Act on Supplying Electricity from Renewables (*Stromeinspeisungsgesetz* - StrEG) entered into force (cf. Bechberger 2000, p. 4). The StrEG provided the basis for a purchase and FIT based remuneration of electricity generated by hydro, wind, solar, landfill and sewage gas as well as biowaste powerplants with a capacity smaller than 5 MWp and a maximum public owner share of 75% by electric utilities (§ 1 StrEG). One of the key innovations introduced was the purchase obligation of renewable electricity (§ 2 StrEG). The FIT received by plant owners depended on a percentage share of the average revenue generated by the utilities per kWh of electricity sold to the consumers (§ 3 StrEG). For solar and wind this percentage was set to 90% and for the other types to 80% until a capacity of 500 kWp and, in case of transgression reduced to 65%. According to Bechberger and Reiche (2004) the support for electricity from PV was too low for a market entry under this scheme, granting a remuneration of approximately 0.085 €/kWh opposed to generation costs of about 0.77 €/kWh. Additionally, critiques were raised by the network operators and utilities that saw themselves confronted with too high remuneration costs and regional disadvantages when supporting renewables (cf. Roßegger 2008, p. 39).

In case of PV, the federal and state governments aimed to enhance technology diffusion by introducing the so called *1000 Roofs Program* between 1991 and 1995, resulting in a cumulative installed PV capacity of 4 MWp (cf. Bechberger and Reiche 2004). After the change of the federal government in 1998 an extended support of renewable energies was envisaged and, for PV, realized by the setup of the *100,000 Roof Program* in 1999 (cf. SPD/Grüne 1998). The program designated a capacity enlargement of PV by 300 MWp funded by a subsidy of 510 million€ with the aims of stimulating private investments by a factor of two to three and establishing Germany as a location for the uprising PV industry (cf. Bechberger and Reiche 2004). However, its great public success appeared only in combination with the issue of the EEG in the year 2000, which provided higher returns on investment because of the implemented FIT scheme (see below) (cf. Bechberger and Reiche 2004).

Because of the liberalization of the European electricity markets in 1998, emission reduction duties from the Kyoto protocol 1997, the governmental change with higher attention on promotion of renewables and inherent structural deficits of the regulation, the StrEG had to be rethought and revised. Especially the electricity price linked FITs were seen critically because the market liberalization had led to a price decrease which was followed by a reduced remuneration for electricity from renewables, making them, related to their high investment cost, even less attractive to potential investors. Moreover the electric utilities were burdened differently according to regional deviations in renewable electricity potential and thus capacity expansion. This led on the one hand to lawsuits against the StrEG and consequently reduced its planning security and, on the other hand, was in conflict with requirements of the electricity market liberalization enforced by the EU. (cf. Bechberger 2000, pp. 8-13)

Prior to the enactment of the EEG as the successor of the StrEG, different research studies were commissioned by the German Federal Ministry of Economics and Technology (BMWi), BMU and the German Federal Environmental Agency (UBA), among others, in order to setup a new or adjust the prevailing promotion scheme for renewable energies. Technology specific subventions, cost oriented FITs, quota solutions and improved burden sharing were discussed. Out these studies the political parties and the BMWi and BMU issued drafts that generally referred to a continuation of an adapted StrEG with a call for cost oriented FITs, purchase obligations and a burden sharing on transmission system operator (TSO) level. However, the details of the drafts differed significantly. In case of PV, for instance, cost oriented FITs between 8.4 and 50.6 $cent/kWh$ (ct/kWh)[5] were proposed. After a phase of political and expert discussion and final adaptions the EEG was adopted by the parliament on the 25th of February 2000. (cf. Bechberger 2000, pp. 59-62)

2.2.2 Vertices and Progress of the EEG's FIT Scheme

The EEG entered into force on April 1st, 2000 and has undergone major amendments in 2004, 2008-2009 and 2012. Subsequently the different versions will be described with a focus on the PV FIT policy design options implemented and the FIT rate evolution. Support details of other renewable energies are not specifically assessed but, in case they could be relevant for the raised issue of this study, briefly mentioned.

The EEG (2000) introduced a technology specific, fixed FIT system for renewable electricity from newly installed facilities with priority purchase

---

[5] Units have been converted from 16.5 $Pfennig/kWh$ and 99 $Pfennig/kWh$. "Ct" refers throughout this study to €cents.

obligation and remuneration by the network operators with the highest proximity to the facility (§ 3 EEG 2000). The rationale behind this choice was to provide a high level of security for investors and planners (cf. Bundesregierung 2002a, p. 3). Under consideration of the law are technologies generating electricity from hydropower, landfill, mine and sewage gas, biomass, geothermal energy, wind power and solar energy (§§ 4-8 EEG 2000). The FIT level was oriented on the specific investment costs (cf. Bundesregierung 2002a, p. 3; Bechberger 2000, pp. 60-63) and set for PV to 50.6 ct/kWh (§ 8 (1) EEG 2000), which was by far the highest compared to the other renewable electricity technologies in the support scheme. Open ground plants were remunerated up to a capacity of 100 kWp (§ 2 EEG 2000). A fixed yearly degression rate of 5%, that came into force with a two year delay, was introduced to account for the learning PV technology and to provide incentives for improvements and innovations (cf. Bundesregierung 2002a, p. 4). In addition § 8 (2) EEG (2000) fixed a capacity expansion cap of 350 MWp, resulting from the *100,000 roofs program*, above which the remuneration claim was omitted. To support other technologies, more differentiated FIT design options were used. Wind power plants for instance received a higher remuneration in the first years after installation which was later on, depending on the plant yield, decreased (*front-loaded stepped tariff*). In case of landfill, mine and sewage gas, biomass, hydropower and geothermal energy the FITs were classified according to plant sizes to represent differences in investment costs. The duration of the guaranteed support was set uniformly to 20 years (§ 9 EEG 2000). The burden sharing mechanism implemented, obliged the TSOs to balance the amount of renewable electricity among each other and collect payments via an average tariff from the electric utilities that in turn could sell the electricity to end customers (§ 11 EEG 2000). In terms of a revision procedure, § 12 EEG (2000) demanded the elaboration of a "progress report" (*Erfahrungsbericht*) of market and cost developments every two years. Based on this report FITs and degression rates should be adapted.

The first major amendment of the EEG was implemented in 2004 following the recommendations of the first progress report from 2002 (cf. Bundesregierung 2002a). In terms of PV, the report stated that from 2004 on, having accomplished the 300 MWp capacity enlargement of the *100,000 roofs program*, the EEG would have the potential of being the only support instrument for the growing PV market. Out of this conclusion the program for soft loans was not extended but, as a compensation, FIT rates were increased (§ 11 EEG 2004) and no capacity expansion cap was fixed. The base FIT was set to 45.7 ct/kWh and a stepped tariff with an additional premium was introduced according to the installation type and the PV plant size. PV plants installed on[6] a building or noise protection wall were remunerated with 57.4 ct/kWh for a size up to 30 kWp and with 54.6 ct/kWh for the capacity greater than 30 kWp. For plants bigger than 100 kWp the FIT was set to 54 ct/kWh.[7] If the PV installation was not part of the roof of a building or the roof itself, e. g. a PV facade solution, the FIT was additionally increased by 5 ct/kWh. For open ground PV plants the basic FIT was granted if it was in the scope of a zoning map (*Bebauungsplan*) or on an area that had undergone a plan approval procedure (*Planfeststellungsverfahren*). It was further specified that those locations only included areas that had already been sealed, were agricultural or military conversion areas or were green areas that had been designated for the purpose of setting up a PV plant in the zoning map and were formerly used as agricultural land. The degression rate stayed at a level of 5% per year starting in 2005 for on-roof facilities. From 2006 on the degression rate increased to 6.5% for open ground plants. Facade PV solutions were excluded from the degression rate.

Moreover, the amendment adapted the burden sharing mechanism by introducing exemptions for industrial and railway companies (§ 16 EEG 2004) and changed the revision procedure (§ 20 EEG 2004) by demanding a next progress report prepared by the BMU in 2007 and from then on every four years.

---

[6] "On" includes both, rooftop and facade PV installation solutions.

[7] Generally the total FIT granted for 1 kWh provided by a specific plant is calculated partially accounting for the corresponding FITs in the respective size classes (§ 12 (2) EEG 2004). Having installed a 10 kWp PV plant would result in a FIT of 57.4 ct/kWh for the first 30 kWp, i. e. 75% and 54.6 ct/kWh for the remaining 10 kWp, i. e. 25%. Hence, on average a FIT of 56.7 ct/kWh would be received. The calculation rationale was maintained until the current version of the law.

The 2007 progress report (cf. BMU 2007) was based on a research report of Staiß et al. (2007) that conducted an in-depth analysis of market and cost structures for the technologies considered in the EEG. Moreover questions of environmental conformity, energy market integration, storage technologies and, again, the burden sharing mechanism were addressed. With the calculation of the levelized cost of electricity (LCOE) measured in ct/kWh for different PV plant sizes and types (on-roof, facade and open ground) it could be conducted that FITs have been sufficient to cover these costs (cf. Staiß et al. 2007, pp. 264-265). The report gave the following recommendations for a revision of EEG's regulation concerning PV: at first, an increase of the degression rate to enforce higher cost savings at the producer level that should be passed on to the consumer or alternatively a onetime lowering of the PV FITs because they were found to be too high in the 2004 version. Secondly, a modification of the stepped tariff design to provide a "close to the market" remuneration via the introduction of plant size categories (smaller than 10 kWp and greater than 1,000 kWp) was proposed. Most interesting for the research questions of this study are the discussions on promoting high quality facilities by an additional premium for those PV plants that can provide certificates for environmental and quality standards or that assure a predetermined recycling rate. These were, however, disliked because of inducing too much complexity in the law (cf. Staiß et al. 2007, p. 278). Net metering possibilities with premium options were discussed as well but not recommended. Within the progress report an additional ecological assessment of the EEG regulation was carried out focusing on the direct environmental effects of open ground PV plants like the impact on birds and mammals, the landscape and possible release of toxic substances (e. g. cadmium from thin-film modules) (cf. BMU 2007, pp. 127-128). The recommendations of the BMU (2007, p. 130) were a onetime lowering of the FITs for roof and open ground plants, an increase of the degression rate for roof and open ground plants and the introduction of a power class greater than 1,000 kWp.

The amendment of the EEG took place in 2008 and entered into force in the beginning of 2009. The law was enlarged from 21 to 66 paragraphs indicating the rising complexity of the FIT scheme and related regulation. The changes comprised detailed regulations to increase the legal security for the priority grid access (§§ 4, 9, 10 EEG 2009) and grid feed-in management regulations (§ 11 EEG 2009), allowing grid operators to control the feed-in amount and secure grid stability. Amendments in the FIT scheme included, among other smaller changes, a register obligation for all plants to improve data quality, a detailed regulation for direct marketing (§ 17 EEG 2009) and an adaption of FITs and degression rates in accordance to the findings of the progress report (cf. BMU 2007).

In case of PV, the FITs and degression rates were altered significantly. New remuneration classes were introduced according to the plant size for roof and facade installations and those installed on a noise protection wall: up to 30 kWp, 43.0 ct/kWh, from 30 to 100 kWp, 40.91 ct/kWh, from 100 kWp to 1 MWp, 39.58 ct/kWh and above $1MWp$, 33 ct/kWh were granted as FITs (§ 33 (1) EEG 2009). For open ground PV systems 31.94 ct/kWh were determined (§ 32 EEG 2009). The degression rates have been increased and coupled to a flexible capacity expansion cap (*atmender Deckel*). Specifically, if the cumulated installed capacity in 2009 were smaller than 1,000 MWp the degression rate would decrease by 1% and if it is exceeding 1,500 MWp the degression rate increases by 1%. Those capacity ranges for degression rate adaption were also defined until 2011. This approach can be classified as an indirect capacity dependent tariff revision (cf. Klein et al. 2008, p. 23). Additionally a possibility of receiving a remuneration of 25.01 ct/kWh for the own use of the produced renewable electricity was created for plants smaller than 30 kWp (§ 33 (2) EEG 2009).

In 2009 the ordinance on a nationwide equalization mechanism (*AusglMechV*) was enforced according to § 64 EEG (2009). It represented an important new regulation with regard to the burden sharing mechanism which is currently still applicable. The ordinance obliged the TSOs to market the EEG-electricity on the spot-market (§ 2 AusglMechV) and thereby released them from the duty of physically transferring the electricity to subsequent utilities (§ 1 AusglMechV 2009). The revenues from the electricity sale at the spot-market (market premium) and the remuneration costs for that electricity (from the FIT scheme) are settled and the difference (*EEG-Differenzkosten*) is divided among the total of remunerated electricity resulting in the reallocation charge (*EEG-Umlage*) that can be passed on to the electric utilities and thus to the end consumers (§ 3 AusglMechV).

Between the 2009 amendment and the EEG (2012) seven revision laws have been issued continuously adapting the EEG. For the purpose of this chapter the focus will be set on the regulations relevant for the PV FIT scheme that were issued in August 2010 (EEG 2010-PV) and April 2011 (EEG 2011) and entered into force in July 2010 and May 2011 respectively. In the EEG (2010-PV) FITs were lowered onetime by 8-13%, depending on the type installation. For plants connected to the grid from October 2010 on the tariffs were decreased additionally by 3%, excluding open ground plants that were connected in the beginning of 2011 and were in the scope of a zoning map. The revision removed FITs for PV plants on agricultural areas but extended the conversion area definition to those areas with traffic and housing use. Thereby PV plants installed next to railways or highways were included in the scheme. Moreover the degression rates were subject to an increase and diversification. Now different capacity expansion corridors were defined entailing corresponding degression rates. In 2011 for instance, the degression rate would increase by 1% if 3,500 MWp capacity expansion were exceeded, 2% if 4,500 MWp were exceeded and so on up to 6,500 MWp and 6%. The degression rate decrease followed the same principle with specifically defined expansion caps. The own-use remuneration introduced in the 2009 version was extended to plants with up to 500 kWp capacity. The premium for own use received was established by the FIT rate reduced by 16.38 ct/kWh for 30% of the plants yield. Above these 30% the FIT is reduced only by 12 ct/kWh giving the operator higher incentives to avoid grid feed-in.

However, with the revision in April 2011 (EEG 2011) the FITs, degression rates and corridors for PV adapted again. The degression rate linking to specific capacity expansion corridors was retained and further increased with the additional possibility of a biannual adaption. The corridors were changed from 3,500 MWp to 7,500 MWp in steps of 1,000 MWp with a fixed annual degression rate of 9% and a flexible rate from 3% to 15% depending on the capacity expansion. For an expansion path lower than anticipated by the law, the dregression rate could also be decreased.

In 2011 the law had undergone another major amendment leading to the EEG (2012) which entered into force at the 1st of January 2012. This action was based on the findings of the progress report 2011 (cf. Bundesregierung 2011). It proposed an extension of the former regulation on direct marketing by introducing an optional, technology specific, gliding market premium scheme with the aim of a better market integration through a demand oriented generation as required by the energy concept of the federal government (cf. BMU 2011, p. 7; Gawel and Purkus 2013). The law amendment implemented the recommendation by giving the plant operator the chance to choose between receiving the FIT or a market premium for electricity directly marketed (§ 33d EEG 2012) with the possibility to split the electricity generated in a share remunerated by the FIT scheme and one remunerated by the sales revenue and the market premium (§ 33f EEG 2012). The duration over which the electricity is directly marketed is added to the FIT payment duration (§ 33e EEG 2012). The market premium received is the difference between the FIT rate and a technology specific reference market value, which corresponds to the difference between monthly average electricity prices observed at the spot-market for the technology under consideration and a fixed but over time decreasing management premium. The latter should account for transaction costs accruing from the direct electricity marketing (cf. Annex 4 EEG 2012). This model can be referred to as a modified version of the spot-market model discussed by Couture and Gagnon (2010) (see above). However, it remained valid only for a short time and was in the next amendment replaced by a market integration model (see below). Within the EEG (2012) the recommendations of the progress report were followed to exclude open ground plant support on conversion areas located in a natural protection area or national park (cf. § 32 (2) EEG 2012 and Bundesregierung 2011, p. 19).

However, because of high costs caused by the massive PV capacity expansion between 2010 and 2011 (14.9 GWp) BMU and BMWi proposed severe cutbacks in the FIT support for PV in February 2012 (cf. BMWi and BMU 2012a, pp. 7-9). The paper suggested to decrease the FITs once in March 2012, to introduce a monthly degression and to remove the premium for own use. Moreover only a share of 85-90% of the generated PV electricity should be subject to remuneration and the expansion corridors should be yearly reduced by 400 MWp starting from 3,500 MWp until they reach 900-1,900 MWp in 2017. The paper resulted in a law draft presented in March 2012 (cf. BT-Drucksache 17/8877) that postponed the changes to April and triggered massive protests by the solar industry, labor unions and environmental organizations (cf. e. g. BSW-Solar 2012). The draft was passed by the parliament but stopped by the federal council. As a consequence a mediation council had to be convened that settled the conflict with a partial relief of the cutbacks at the end of June 2012 (cf. BMU 2012b). The law amendment was issued in mid August 2012, entered into force retroactively by April 2012 and represents the current version of the EEG (EEG 2012-PV). Implemented major changes are the introduction of a total cumulative capacity cap for PV of 52 GWp (§ 20b (9a) EEG 2012-PV) beyond which the support is completely ceased out,[8] a removal of the additional premium for own use, the setup of a market integration model and adaptions in FIT and degression rates as well as in PV plant size classifications.

The FITs for plants installed on buildings are now granted according to the following size dependent classification: up to 10 kWp (19.5 ct/kWh), between 10 and 40 kWp (18.5 ct/kWh), between 40 kWp and 1 MWp (16.5 ct/kWh) and finally between 1 and 10 MWp (13.5 ct/kWh). The capacity share of PV plants greater than $10MWp$ receives no FIT payments. Open ground plants receive 13.5 ct/kWh until a capacity of 10 MWp. The amendment introduced a fixed capacity expansion corridor lying between 2,500 and $3,000MWp$ per year (§ 20a EEG 2012-PV). A monthly degression rate is introduced, 1% until November 2012 and from then on linked to the capacity expansion extrapolated for one year. When-

---

[8] Priority grid access is not affected in case of an transgression.

ever the corridor is exceeded by a certain capacity (five corridors from 1,000 MWp to 4,000 MWp) the degression rate is increased additionally by 0.4 to 1.8%-points (§ 20b EEG 2012-PV). The market integration model constrains the remunerable amount of generated electricity to 90%. The scheme is applicable for the capacity share of PV plants installed on buildings and noise protection walls between 10 kWp and 1 MWp (§ 33 EEG 2012-PV). Consequently, the generated electricity attributable to the first 10 kWp and the one exceeding 1 MWp receive full remuneration. The Bundesregierung (2012) states that thereby operators of PV plants should receive incentives to plan their facility in line with market conditions and consume or market a part of the generated electricity by themselves. Payments received for the non-remunerable share are in this case the revenues gained at the spot-market for PV electricity (§ 33 (2) EEG 2012-PV).

2.2.3 Goal Development

After having presented the development of the EEG's FIT scheme for PV and the current legislation the evolution of the law's goals in its different versions is addressed. The environmental goals of the EEG are of particular interest for this study since they could provide a potential justification and legitimization of the carried out research in the upcoming chapters.

The EEG (2000) defines its goal as enabling a sustainable development in energy supply with regards to climate change on the one side and environmental protection on the other. Furthermore, the share of renewable energies in electricity supply should be significantly increased in order to accomplish the aims set by Germany and the EU to at least double the share of renewables in end energy supply until the year 2010 (§ 1 EEG 2000). Not explicitly mentioned in the law but a reasoning behind its introduction was the intention to promote a dynamic development of different technologies deploying renewable energy sources to enable competitiveness with conventional ones in the mid- and long-term and to strengthen the position of the German industry in the global market (cf. BT-Drucksache 14/2341).

With the first amendment in 2004 the list of goals was adapted. In line with the promotion of a sustainable development of the energy supply with regards to climate and environmental protection the term "nature protection" was added. Besides the endeavor for supporting sustainable development, the aim of decreasing the social cost of energy supply, considering the internalization of external effects, was introduced. Moreover, the law should explicitly contribute to avoid usage conflicts of fossil energy sources and promote the advancement of renewable electricity technologies (§ 1 (1) EEG 2004). In section two of § 1 EEG (2004) the formerly only normative goal of doubling end energy from renewables was now specified and quantified. Until 2010 the share of renewables in electricity supply should reach at least 12.5% and was aimed to increase to 20% by 2020. The amendment has been implemented in accordance with the sustainability strategy of the Bundesregierung (2002b) and the European directive 2001/77/EC. In the law explanation the legislator clarifies that the aims outlined in § 1 EEG 2004 are not equivalent but ordered according to their importance with the exception of climate, natural and environmental protection (cf. EEG-Explanation 2004, p. 12). It is also explicitly mentioned that not the renewable capacity expansion itself but its contribution to climate, environmental and natural protection are the purpose of the law. The aim of supporting the development of renewable electricity technologies provides the basis for the FIT design. With the promotion technical and economic innovations that would result in lower social cost and improved environmental performance are envisaged (cf. EEG-Explanation 2004, p. 14). For this reason the EEG provides technology specific FITs with a degressive component that should give incentives for such improvements and innovations, so that renewable energy technologies can become competitive (EEG-Explanation 2004, p. 14).

The amendment in 2009 changed § 1, as a result of G8-Summit and decision of the EU council in 2007 (cf. EEG-Explanation 2009, p. 19). The aim of natural protection was removed because of a changed understanding of the environment in which the term "nature" is already included. However, with this shortening of § 1 EEG 2009 the legislator was not intending to decrease the importance of natural protection (cf. EEG-Explanation 2009, p. 19). The 2009 amendment also removed the, in 2004, newly introduced objective of avoiding usage conflicts to the favor of the goal to spare fossil energy resources (§ 1 (1) EEG 2009). The term was adapted to account for broader consequences of fossil fuel use comprising import dependency, responsibility towards future generations and the avoidance of usage conflicts (cf. EEG-Explanation 2009, p. 21). In section two of § 1 EEG 2009 the quantitative objective of reaching a 20% share of renewables in electricity supply was increased to 30%. The long-term orientation should additionally create investment and planning security (cf. EEG-Explanation 2009, p. 22).

The last major amendment in 2012 did not modify section one of § 1 but refined the long-term expansion goals of section two. The legislator thereby included the objectives of the German energy concept (cf. BMU 2011) in the legal text. It is envisaged to accomplish a share of 35%, 50%, 65% and 80% in the years 2020, 2030, 2040 and 2050 respectively. Most importantly, the requirement of the successful integration of renewable electricity in the overall electricity system is included since this discussion was specifically highlighted in the 2011 progress report (cf. Bundesregierung 2011, pp. 10-12) and intensively analyzed in the accompanying research report of Sensfuß (2011b, pp. 117-185). Moreover a third section was added in the legal text. It emphasizes that the EEG is additionally a mean to accomplish the aim of reaching a share of 18% renewables in gross end energy provision as requested by the European directive 2009/28/EC (cf. EEG-Explantion 2012, p. 119).

Tab. 2.2 summarizes the goal development within the EEG from the date of implementation in 2000 until its last major amendment in 2012. In line with the already identified rising complexity of the regulation, the goals became more diversified in order to cover a broader range of energy policy issues and comply with other German and EU legislation.

Tab. 2.2: EEG goal development; Source: Based on EEG (2000, 2004, 2009, 2012-PV)

Goal		2000	2004	2009	2012
Climate change		x	x	x	x
Environmental Protection		x	x	x	x
Nature Protection			x		
Decrease costs of electricity supply			x	x	x
Avoid usage conflicts of fossil energy sources			x	$(x)^3$	$(x)^3$
Fossil energy resources conservation				x	x
Promote advancement of renewable technologies		$(x)^1$	x	x	x
Share of renewables in electricity supply	2010	$12.5\%^2$	12.5%	-	-
	2020	-	20%	30%	35%
	2030	-	-	-	50%
	2040	-	-	-	65%
	2050	-	-	-	80%

[1] Not specifically mentioned in the legal text but a basic reason for the EEG's introduction
[2] Not specifically mentioned in the legal text but the corresponding share according to Bechberger and Reiche (2004)
[3] Aim is a component of the aim to spare fossil energy resources

2.3 Reasoning behind the EEG's PV FIT scheme

This section aims at giving insight into the mechanism applied in order to determine the level and duration of FIT support in the EEG. Since this study aims at assessing this FIT scheme according to environmental criteria it is crucial to capture the rationale of the legislator in the determination of the tariffs.

By setting the tariff level the legislator has to ensure that it is sufficiently high for giving investment incentives in the renewable energy technology and, at the same time, to avoid an overcompensation of the recipient (cf. Mendonca et al. 2010, p. 19). Because of this, the remuneration level is considered an important determinant for policy effectiveness and accruing support costs (cf. IEA 2008, p. 90). According to Menconca et al. (2010, p. 19) there are different possibilities to determine the level of support, either by setting the tariff in accordance with electricity generation costs resulting from the renewable technology, based on avoided costs of conventional electricity supply, avoided external costs from generating renewable electricity or by linking the FIT to the electricity price. However, the option of orienting the FIT on the electricity generation costs is discussed as a best practice because it would offer a predictable rate of return for the investor and could easily account for differences in renewable technologies (Mendonca et al. 2010, p. 19). By not basing the FITs on technology specific generation costs there is the risk that either plant operators do not receive sufficient money for the investment amortization or that the tariff level remains too high (cf. Mendonca et al. 2010, p. 62). In this sense, Madlener and Stagl (2005) add that, even though electricity generation cost are a yardstick for the FIT determination, in practice national interests and lobbying skills also play a role. However, according to Klein et al. (2008, p. 11) and Mendonca et al. (2010, p. 20) the following determinants of electricity generation costs should be considered in the FIT level determination:

- Investment for the plant
- Other project costs
- Operation and maintenance costs
- Fuel costs when applicable

- Inflation
- Interest rates
- Profit margins for investors
- Decommissioning costs
  when applicable

Having this data raised, an estimation of the expected power yield and the duration of support fixed for the specific technology enable to calculate the FIT level by applying dynamic investment calculation methods (cf. Menonca et al. 2010, p. 20). Klein et al. (2008, p. 25) argue that the differentiation of FITs based on technologies might not be sufficient because generation costs are also influenced by plant size or location dependent characteristics (e. g. solar radiation). Large plants may require a lower upfront investment per installed unit of capacity than small plants because of economies of scale. Concerning the location, it has to be considered that the higher the electricity yield of a plant the lower will be the LCOE. Therefore a differentiation via a stepped tariff design (see above) could be reasonable to adjust the FIT level to special conditions (cf. Klein et al. 2008, p. 27). The duration of the FIT support is in addition inherently linked to the support level. Shorter periods would require higher FITs to amortize the initial investment and granting support over a longer time could reduce the specific remuneration for one kWh fed into the grid. Del Río (2012) argues that long periods would reduce the investment risk, provide stability and trigger R&D investments and learning effects, but could adversely lead to consumer lock-ins, high follow-up costs and reduced technological competition.

In the context of the EEG, FIT rates are determined on the basis of renewable energy technology specific investment, operation and capital costs related to the average lifetime of a technology and a market oriented interest rate (cf. BT-Drucksache 14/2341). By applying this scheme for FIT level determination renewable technologies should become competitive with other forms of electricity generation and an environment with high planning and investment security should be created (cf. Bundesregierung 2011, p. 7; 2002a, p. 3). As it has been shown above plant size specific stepped tariffs have been applied since 2004 to account for economies of scale in LCOE (cf. Staiß et al. 2007, p. 277; Klein et al. 2008, p. 77) and the duration has been set uniformly to 20 years.

The determination of FIT levels for PV in particular is carried out in the progress and accompanying research reports of the EEG. In these reports market and cost developments are reviewed and LCOE from PV are determined for a representative set of exemplary plants with different characteristics (cf. Staiß et al. 2007, pp. 262-265; Reichmuth 2011, pp. 106-118). Reichmuth (2011, pp. 106-107) for instance defined five different cases from which three are roof mounted (5, 30 and 1,000 kWp with Mono- and Multi-Si PV modules) and two are open ground PV plants (1 MWp and 20 MWp with thin-film cadmium telluride (CdTe) modules). A yield of 900-950 kWh/kWp is assumed.

The methodology used for deriving the LCOE of these model cases is outlined by BMU (2007, p. 167-175) and Schmidt (2011, pp. 6-8) and based on a dynamic annuity calculation, according to the norm

2067:2012 of the Association of German Engineers (VDI 2067:2012, current version), in nominal terms over the time of remuneration (20 years) considering an inflation rate of 2% per year. The interest rate is derived as a weighted sum of the observed interest on equity and borrowed capital, with the weights being the share of observed equity or loan financed projects. This calculative interest rate should thereby also represent an expected rate of return for the investor. Annuities for investment, fuel, operation and other costs or revenues are considered in the appraisal. The LCOE are derived from the annual costs divided by the expected annual electricity yield (e. g. BMU 2007, p. 172). The LCOE can be illustrated by Formula (2.1) (based on Schmidt 2012):

$$LCOE = \frac{\frac{\sum_{t=0}^{T}(I_t + OM_t)}{(1+r)^t} * \frac{r(1+r)^T}{(1+r)^T - 1}}{ASAY} \qquad (2.1)$$

where I_t are the investments made in the year t, OM_t the operation and maintenance cost in the respective year, $(1 + r)$ the assumed discount factor and $ASAY$ the average specific annual electricity yield. The term $\frac{\sum_{t=0}^{T}(I_t + OM_t)}{(1+r)^t}$ represents the present value of all life cycle costs and $\frac{r(1+r)^T}{(1+r)^T - 1}$ represents the annuity factor for a period T and an interest rate r. The product of both terms can be interpreted as the annuity of life cycle electricity generation cost (cf. Schmidt 2012).

For PV, Reichmuth (2011, p. 110) used a calculative interest rate of 5% per year. Out of the calculation, LCOE ranging from 18 to 27 ct/kWh could be derived, being highest for the small roof mounted plants and lowest for the large scale open ground plant deploying thin-film modules (cf. Reichmuth 2011, p. 188). Compared to the, at the time valid, FITs it could be observed that they all cover the LCOE. In the 2007 progress report a more diversified portfolio of PV plants was assessed (cf. BMU 2007, p. 126). It comprised three roof mounted, two facade mounted and two open ground plants with different sizes. For these plants nominal LCOE were found to range from approximately 38 to 78 ct/kWh, with the lowest LCOE achieved by the open ground systems and the highest by a facade system. Thereby the enormous cost decrease in PV systems can be shown. The costs for generating electricity of a small 5 kWp roof mounted plant for instance changed over the time in which the two progress reports were issued (2007 and 2011) by 50%. This is why Schmidt (2012) highlights the necessity for a regular reviewing of the FITs in comparison to the cost performance.

This decrease in cost is represented and partly also generated by the introduction of degression rates in the FIT scheme. According to Staiß et al. (2007, p. 275) the degressive tariff was introduced in the EEG to influence the investment costs for PV plants by giving an incentive to the producers to pass on cost savings in the production process to the end customers. Under the assumption that the FIT rate in the respective year fully represents the generation costs and all actors are informed about and can anticipate the degression rate, a FIT degression would deteriorate the profitability of the investment in the renewable technology resulting in a shrinking demand that ultimately leads to a decrease in price (cf. Reichmuth 2011, p. 210; Staiß et al. 2007, p. 275). For this reason it is of great importance to (1) represent the current LCOE for the technology in the FIT and (2) determine degression rates that correctly estimate future innovation potentials. Concerning the second aspect, Reichmuth (2011, pp. 209-210) argues that a too high degression rate can lead to a market collapse and too low rates can create windfall profits in the supply chain or at the end consumer level, increasing the total costs for society. Because of this dependency the degression rates are used by the legislator as a tool to control the costs of the support system. Higher degression rates can reduce the accruing support costs, defined as "differential costs" arising from the total remunerations paid, including FITs and direct marketing premiums, and the revenues of the electricity sale at the spot-market[9] (cf. Reichmuth 2011, p. 210).

Concerning the relation of PV systems installed on buildings or noise protection walls and open ground plants, between which the EEG differentiates since its introduction in 2000, the legislator states that generally the former should be preferred over the latter and that this prioritization should also be reflected in the FIT level (cf. EEG-Explanation 2004,

---

[9] Plus avoided grid usage charges.

p. 42; 2009, p. 64). However, the magnitude of this prioritization remains unaddressed by the research and progress reports. An indication is given by Staiß et al. (2007, p. 277), who state that the remuneration of large roof mounted system (from 1 MWp) should be oriented on the FIT for open ground plants but receive some kind of premium to reflect the superiority of using roof instead of open ground area. Thus, it can be assumed that those systems entail equivalent costs but are remunerated differently. Comparing the FIT level of a 1 MWp PV plant to an 1 MWp open ground system under the framework conditions of the EEG (2009) (01.01.2009), a difference of 7.84 ct/kWh or approximately 25% can be identified, pointing at a significant bonus for the roof installed system. From a technical point of view this bonus would then be granted on top of the LCOE for roof systems previously determined according to Formula (2.1). Unfortunately, no reliable data on the extend of this bonus is available. It is also interesting to observe that Staiß et al. (2007, p. 264) and Reichmuth (2011, p. 118) do not address this aspect in their economic analysis of the PV FIT scheme and that the difference between the identified LCOE of roof installed systems and the corresponding FIT level is not necessarily higher than this difference in case of open ground systems.

2.4 Effects and Induced Costs

In 2011 renewables contributed with 20.3% to the gross electricity consumption (cf. BMU 2012a). Compared to the year 2000 this is an increase of 13.5%-points. In total approximately 123 GWh of renewable electricity were generated in 2011 which could avoid about 86 $GtCO_2eq$ emissions that would have been otherwise emitted by the average electricity mix (cf. BMU 2012c, p. 14). Fig. 2.1 depicts the development of the GHG avoidance induced by deploying renewables in Germany on an annual basis. For PV, the values have been estimated based on average emission avoidance factors and the amount of generated PV electricity in the respective year as stated by *Arbeitsgemeinschaft Erneuerbare Energien*

(AGEE 2012). Hence, the increased deployment of renewables leads, *ceteris paribus*, to higher emission avoidance.[10]

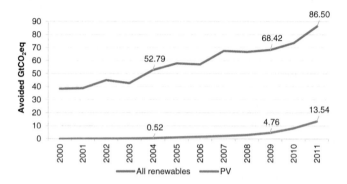

Fig. 2.1: Avoided $GtCO_2eq$ by deploying renewables in Germany; Source: Authors based on AGEE (2012)

The installed PV capacity showed the highest increase among the supported technologies. It was enlarged from the year 2000 until 2011 by approximately 24.8 GWp, from which 14.9 GWp were installed between 2010 and 2011 (cf. REN21 2012, p. 101). Thereby Germany became the worldwide biggest market for PV systems and deploys the largest PV operating capacity (cf. REN21 2012, p. 48). Fig. 2.1 shows that in 2011 around 13.5 $GtCO_2eq$ could be avoided by the PV electricity production. The most recent figures on the installed PV capacity under EEG remuneration are published by the *Bundesnetzagentur* and show a sustained growth of the German PV market. Despite the restrictive changes in the EEG (2012-PV) 7.6 GWp have been installed in 2012 and, in January 2013, 274.67 MWp were added so that a total operating capacity of 32.66 GWp is currently remunerated by the EEG (cf. Bundesnetzagentur 2013). The capacity expansion corridors defined by the EEG (2012-PV) were exceeded resulting in a monthly decrease of the FIT for PV plants

---

[10] The reader is encouraged to notice that there is a lively discussion about the net emission avoidance from renewables when considering interdependencies with other GHG mitigation policies like the European ETS (cf. e. g. Lehmann and Gawel 2013; Gawel et al. 2013b; UBA 2011 or Frondel et al. 2008, 2010).

by 2.2% from February 2013 on. Based on this, the current FIT level (March 2013) for new installations ranges between 11 and 16 ct/kWh depending on the plant size.

Fig. 2.2 depicts the size specific development of PV capacity installations from the year 2000 until January 2013. A strong trend towards the installation of larger PV plants can be identified. Until the 2004 amendment of the law small residential plants contributed the most to the annual newly installed capacity and large scale plants greater 1 MWp were not installed at all. From 2004 on the share of large PV plants increased rapidly (6%, 2004 to 42%, 2012). This soaring can be explained by the removal of the 100 kWp cap for open ground plants (cf. Reichmuth 2011, p. 13). Since 2009 small plants only play a marginal role in yearly capacity adding. The 2013 data cannot be considered representative because it only accounts for one month. Since the EEG prefers the deployment of roof areas over open ground areas (cf. Reichmuth 2011, p. 257) the 2012 revision (EEG 2012-PV) tries to constrain the observable trend by excluding the remuneration of capacity shares greater than 10 MWp. If this measure proves to be fruitful will become apparent at the end of this year.

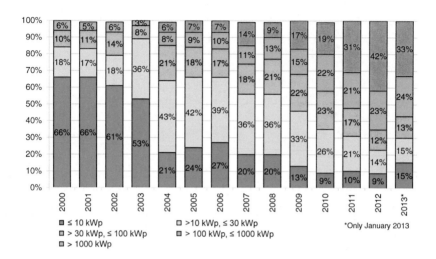

Fig. 2.2: Shares of PV capacity adding in different size segments; Source: Authors, 2000-2010 data based on Reichmuth (2011, p. 13), 2011-2013 data based on Bundesnetzagentur (2013)

Because of the high deployment rates of renewables the EEG is generally considered as a great success (cf. Bundesregierung 2011, p. 3 and Reichmuth 2011, p. 214). However, the high capacity enlargements also come at a cost. The EEG differential costs and the corresponding reallocation charge rise steadily and thereby burden the electricity consumer. To counteract this development the progress report 2011 already aimed at improving the cost-efficiency and limit overcompensation in order to secure the financing mechanism of the law (cf. Bundesregierung 2011, pp. 7-8). The amendment in 2012 (EEG 2012) therefore removed the own use premium regulation that was found to have overall negative consequences (cf. Reichmuth 2011, pp. 167-168) and the revision in mid 2012 (EEG 2012-PV) additionally introduced the capacity cap for remuneration of 52 GWp and the market integration model. The latter aims at significantly decreasing the differential cost from PV (cf. BMWi and BMU 2012b, p. 41). Despite the efforts in amending the law in order to limit the total costs, the total EEG differential costs amounted to ap-

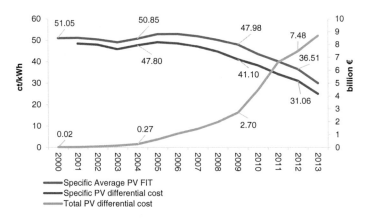

Fig. 2.3: Nominal PV average FIT rate and differential cost development;
Source: Authors based on BDEW (2013, pp. 37-40)

proximately 12.9 billion€ in 2012 and are expected to rise to 16 billion€ in 2013 (cf. BDEW 2013, pp. 37-38). This cost evolution resulted in an increase of the reallocation charge from around 3.6 ct/kWh in 2012 to 5.3 ct/kWh in 2013. The share of the reallocation charge attributable to PV installations amounts to 56.1% and 53.4% respectively.

Fig. 2.3 depicts the cost development of the EEG PV support in nominal terms. The specific differential cost for 1 kWh of PV electricity (red) and the average PV FIT rate paid (blue) are measured on the left, the total PV differential costs (green) on the right axis. Highlighted are the values of each graph at the time of the EEG's major amendments. It can be seen that from 2004 on, when the FIT rates for new plants were increased, the specific differential cost and average FIT payments rose and from 2006 on started to fall again because of the incipient degression. It has to be noted that the average FIT payments refer to all EEG FIT expenditures for PV in a specific year divided by the total PV electricity generation. Therefore, all PV plants operated under the EEG in the respective year are accounted for. This is why the FIT levels from Fig. 2.3 significantly deviate from the ones for new plants in the same year. The high capacity adding in the years 2010-2012 (22.5 GWp) in combination with the restrictive governmental measures of increased degression rates and reduced FIT levels from 2009 on show their first successes. In 2009 the average FIT level was still 48 ct/kWh and could be reduced until 2012 by approximately 12 ct/kWh. For 2013 a further decrease by 6 ct/kWh, without including market and management premium and direct marketing options is estimated (cf. BDEW 2013, p. 38). Despite the falling trend of specific cost it has to be acknowledged that the enormous capacity expansion is associated with an exponential rise in total costs that have to be defrayed by the electricity consumer via the reallocation charge. For PV only, the total differential cost rose to 7.48 billion€ in 2012 (see Fig. 2.3). However, the introduction of the expansion corridor in 2009 and the following amendment in 2011 can be seen as a first success in total cost containment since the exponential trend could be interrupted. Conversely, it should be considered that the promotion of renewables can also save external costs by avoiding environmental damage (cf. BMU 2012c, p. 50). The BMU (2012c, p. 56) quantifies the utility from supporting renewable electricity generation to about 8 billion€ for 2011 which are opposed to total EEG differential of 12.1 billion€. However, it is stated that the uncertainty in monetizing avoided environmental impacts are considerably high and that important benefits could not be accounted for (cf. BMU 2012c, p. 56).

In the near future further revisions of the EEG and its remuneration scheme are envisaged to secure the affordability of power and increase the cost-efficiency and security of supply of the promotion scheme (cf. Altmaier 2013; Altmaier 2012; Bundesregierung 2011, pp. 7-8). The federal minister for the environment, Altmaier (2013), for instance, recently proposed to cap the reallocation charge at a maximum increase of 2.5% per year to secure affordability of power for consumers. The discussion also brought up claims to reduce the exemptions from the burden sharing mechanism (cf. Altmaier 2013) since these, at the time, reduced the apportionable amount of electricity by 27% (cf. BMWi and BMU 2012b, p. 36). The TSOs expect a cumulative renewable capacity of 111.39 GW with an electricity generation of 202.96 GWh and related payments to plant operators of 25.39 billion€ until 2017 (cf. TSO 2012). In the case of

PV, the average remuneration is expected to fall from 40.16 ct/kWh in 2011 (cf. BDEW 2013, p. 53) to 25.7 ct/kWh in 2017 (cf. TSO 2012). The effect is mainly attributable to the amended FIT and degression rates for newly installed plants. Because of progressively decreasing LCOE of renewables, increasing prices of electricity from conventional sources and the expected increase in CO_2 certificate prices, the UBA (2011, p. 10) expects the support costs to peak around 2015. Grid parities, or a competitive electricity generation, of renewables are expected to occur for large scale PV plants in 2021 and smaller plants in 2030 (UBA 2011, p. 11).

2.5 Interim Conclusion

From the in-depth analysis of the EEG PV FIT scheme it can be concluded that it comprises a variety of the in Section 2.1.4 identified FIT design options and has undergone a dynamic development which steadily increased the scheme's complexity. This is partly due to the need for legal security, dynamic technology development and new challenges that arise with a higher share of renewables in the electricity mix (e. g. market integration) but also to the need for total cost containment caused by the generous support of renewable electricity technologies.

Increasing the cost-efficiency of the support scheme has recently been a major topic of investigation and reason for amendments justified by the exponential rise in EEG differential cost and the reallocation charge (cf. Altmaier 2013; BMWi and BMU 2012a, pp. 7-9; BMWi and BMU 2012b, pp. 35-39; Altmaier 2012; Reichmuth 2011, p. 304; Bundesregierung 2011, pp. 7-8; Bundesregierung 2010, pp. 8-9).[11] Paradoxically, while cost-efficiency received a high interest the environmental efficiency of the law was only partially covered in the EEG's progress reports and subsequent amendments, although the climate, environmental and resource aims have been a crucial part of the law since its introduction in

---

[11] The references only represent the official debate about the EEG cost-effectiveness. The topic also receives high interest in a controversial academic discussion (cf. e. g. Mennel 2012; Frondel et al. 2008, 2010; Gawel et al. 2012a, 2012b).

2000 (see Section 2.2.3). Madlener and Stagl (2005) support this argument on a more general basis. They argue that renewable promotion schemes usually do not account for adverse environmental, social and economic impacts and rather focus on the short-term economic efficiency. However, environmental impacts of renewable energies have also to be taken into account in order to make regulation decisions based on full cost transparency. Similarly we have to keep in mind the social cost of conventional energy sources (beside climate effects) to make sure that we compare full social cost when arguing the optimal technology mix for energy supply (Gawel et al. 2012a, 2012b). Hence, we need an additional criterion for renewable energy sources support that might be called "environmental efficiency".

Environmental efficiency is understood in this study as the achievement of a predetermined state with the least negative environmental consequences (minimum principle) or *vice versa* a maximized output with the same environmental consequences (maximum principle).[12] In terms of the EEG the minimum principle would for instance lead to the question if capacity expansion goals (e. g. § 1 (2) EEG 2012-PV) have been or will be accomplished with least negative climate, environmental and resource impacts. The terms "environmental criteria" and "environmental efficiency" refer throughout this study to the environmental issues linked to climate change, resource consumption and other detrimental environmental effects.

The FIT level was oriented on the LCOE of a specific technology since the EEG (2000) in order to provide a remuneration that allows competitiveness with other forms of electricity generation (see Section 2.3). This rationale was derived from the goal of the law to promote the advancement of renewable technologies (see Tab. 2.2) which was in turn pursued to increase environmental performance of those technologies and lower social costs. The legislator stated that the objectives outlined in § 1 of the EEG are not equivalent but successive (cf. EEG-Explanation 2004,

---

[12] It should not be confused with the term "eco-efficiency" commonly defined as "*a management philosophy which encourages business to search for environmental improvements that yield parallel economic benefits*" (WBCSD 2000), i. e. creating more economic value with less environmental impacts.

p. 12). Hence, it is questionable why the prioritized goals of climate, environmental and resource protection were not explicitly addressed as an inherent component of the FIT design and their efficient achievement received rather low attention. An explanation for this behavior could be the presumption that enhancing the share of renewables would lead to environmental and climate protection in any case because of the superior performance of these technologies in comparison to conventional ones. With renewables becoming a major part of the electricity system this assumption should be critically scrutinized according to the question of efficiency raised above.

In 2004 the amendment indeed introduced the reporting obligation for an environmental evaluation of the EEG. The law refers to impacts on nature and landscape resulting from renewable electricity plants. Because of this, the progress report of 2007 (and 2011) focused on analyzing the environmental consequences accruing from goal conflicts within the EEG, presuming that the aims of climate change and resource conservation are satisfied with the mere expansion of the operating renewable capacity. Staiß et al. (2007, p. 282-284) address the positive aspects of climate change mitigation and resource conservation on the nature/environment and the possibility of direct negative environmental impacts of EEG remunerated plants. In case of PV the environmental analysis only assesses direct impacts of open ground plants comprising the quality of land for installation, impacts on birds and their habitat, habitat destruction, landscape, soil compaction, distortion from reflexions and magnetic fields and the direct release of contaminants (cf. Reichmuth 2011 pp. 231-293; Staiß et al. 2007, p. 341). From these aspects only the quality of land has been considered in the stepped tariff design for open ground PV. The fact that different forms of installation and PV technologies entail different effects on climate change, resource consumption, amount of space used and other environmental impacts along their life-cycle (see Chapter 3) is not specifically addressed. Accounting for it could have probably decreased the amount of environmental impacts from the same amount of renewable energy generated under the PV FIT scheme. Conversely, in case of biomass for instance, a differentiated system that grants a technology and fuel bonus was already introduced in 2004 (§ 8 EEG, 2004).[13] The differentiation aimed at promoting especially energy efficient and therefore environmental and climate friendly technologies (cf. EEG-Explanation 2004, p. 35) and would therefore contribute to reach the climate protection and resource conservation goals of the EEG more efficiently. As mentioned above Staiß et al. (2007, p. 278) briefly discussed the option of granting a bonus for high quality PV products very briefly, but it unfortunately received no further attention.

Because of the raised shortcomings, namely insufficient assessment of the EEG's environmental goals according to their efficient achievement and deficiencies in addressing such aspects in the FIT design, the following chapters of this study aim at developing possible environmental assessment criteria for different PV technologies and installation types (Chapter 3) and compare them to the observed FIT rates of the EEG (Chapter 4). Thereby it will be assessed if the EEG coincidentally reflects environmental criteria and constitutes an environmental efficient remuneration system.

---

[13] Sustainability criteria in biomass use also receive high interest in EU legislation e. g. in Article 17 and 18 of directive 2009/28/EC.

3 Development of Environmental Assessment Indicators

This chapter presents the deduction process of environmental assessment indicators for PV serving as a representation of a set of criteria derived from the EEG goal analysis of the preceding chapter. Section 3.1 explains the selection process of criteria and indicators for the subsequent analysis and consequently tackles the first sub-research question of this survey. The environmental indicators are based on a previously elaborated LCA model (cf. Töpfer 2012), which assessed the environmental performance of eight different PV installation alternatives deploying crystalline silicon (c-Si) and thin-film PV module technologies. Because of the confidentiality of data contained this chapter summarizes the vertices of the LCA and presents the updated indicator results (Section 3.2). Moreover, a sensitivity analysis on the obtained indicator results is provided (Section 3.4) and the derived findings are compared to those of other authors (Section 3.5).

3.1 Criteria and Indicators for the Analysis

In the preceding chapter it was concluded that there is hardly any attempt within the EEG's FIT scheme development to account for environmental efficiency. Therefore, this section aims at deriving criteria and indicators that can be applied to assess this efficiency for PV. The analysis in this and the following chapters is based on two premises: (1) different installation options and technologies for generating electricity from solar energy using the photoelectric effect do entail different environmental effects stemming from diverging materials used, production processes, places of production and electricity yields and (2) a remuneration scheme that represents such differences would give incentives for investing (more) in the environmentally superior alternative which in turn would result in higher environmental efficiency of the support scheme in total.

To address the identified shortcomings the criteria exemplary assessed according to environmental efficiency are related to the major environmentally oriented goals of the EEG identified in Section 2.2.3: (1) climate protection, (2) environmental protection and (3) fossil energy resource conservation (see Tab. 2.2) and are termed (1) climate change, (2) climate change linked to space requirements and (3) primary resource consumption. They are operationalized by measurable indicators. *"An indicator is a measure, generally quantitative, that can be used to illustrate and communicate complex phenomena simply, including trends and progress over time"* (EEA 2005, p. 7). Communication is the main function of environmental indicators and can only be successful if the indicator sufficiently simplifies the complex reality it tries to represent (cf. EEA 1999). The European Environmental Agency (EEA 1999) further specifies that, especially in the political sphere, indicators are useful to provide information about the seriousness of the raised issue, to support priority setting and to monitor the effects of policy responses. These dimensions shall be addressed by the indicators developed below, by raising awareness about the possibility to address environmental efficiency in the support scheme and refocus on already prioritized but almost unattended goals. Additionally, once such indicators were consolidated they could also be deployed for a monitoring process. However, it should be clearly stated that the chosen indicators in this study are only an example of various conceivable possibilities to represent the climate and environmental protection as well as resource conservation goals of the EEG and could certainly be replaced by others or further developed.

Environmental indicators can be classified within the Driver- Pressure-State-Impact-Response (DPSIR) model that develops a simplified causal chain of how human activities (drivers) exert pressures on the environment, how these alter the environmental state, lead to impacts on environmental quality and how the socio-economic system responds to this change (cf. EEA 1999). The indicators developed in this study are based on an LCA comprising all life cycle stages of a set of PV systems. The PV LCA model can therefore be considered as a conceptual extract of the economic system that constitutes the drivers in the DPSIR model (cf. UNEP 2010, pp. 17-19). Human activities in each life cycle stage (e. g. resource extraction, processing, manufacturing, transport, use, disposal etc.) exert pressures on the environment that can lead to a changing environmental state with adverse impacts (cf. UNEP 2010, p. 19). Thus, indicators directly derived from the LCIA generally refer to the processes of transforming environmental pressures into impacts. LCA literature therefore differentiates between so called midpoint and endpoint indicators in the LCIA (cf. Goedkoop et al. 2009, p. 1), where material balances of the product model (i. e. LCI) are translated into specific potential impacts through a classification and characterization process (cf. ECJRC 2010a, p. 275). Midpoint indicators generally represent a part of the causal chain between pressure release and damage (adverse impact) on the environment. They are most commonly used for criteria originating from a variety of sources but, once transformed into pressures, show an equivalent behavior (cf. ECJRC 2010b, p. 4). This is for instance the case for climate change. GHG emissions occur from a variety of human activities but, once they are released to the atmosphere, they have comparable impacts on the climate system. On the other hand, endpoint indicators are damage oriented and aim at describing the resulting impact on the environment (e. g. impact of toxic substance on ecosystems or biodiversity) (cf. ECJRC 2010b, p. 4).

Niemeijer and de Groot (2008) mention that, depending on the pursued target of analysis, a focus within the DPSIR framework is necessary. They for instance recommend to focus on impact and state indicators if the seriousness of a problem shall be examined. If, on the other hand, the objective of analysis were the control of a situation, pressure and response indicators might be more applicable. Generally both of these foci could suit the present analysis since either the potential direct impact on the environment (damage) or the pressure that will lead to changes in environmental state could be the reference for assessing the effects of PV electricity generation. The first aspect could be represented by LCA endpoint indicators and the latter by a set of midpoint indicators. The choice in this survey is made in favor of the midpoint indicators, representing pressures instead of damages on the environment, because of impacts are more diverse and complex and, at least in LCA methodology, possibly include weightings as well as normalizations of midpoint indicators to "typical" values (cf. Goedkoop et al. 2009, pp. 2-3). Moreover, it is possible to circumvent the decision if an environmental or human centered point of view is chosen for analysis (cf. Niemeijer and de Groot 2008). This aspect is educible by the example of climate change: the emission of GHGs entails a rise in temperature and the temperature increase is a threat to human health and ecosystems, which are the respective endpoint measurements (cf. Goedkoop et al. 2009, p. 23). Focusing only on the emission of GHG avoids further uncertainties and assumptions in the impact assessment.

Hence, indicators derived in this chapter are further transformations of descriptive midpoint LCIA indicators. They relate drivers to environmental pressures and could therefore be interpreted as efficiency indicators (cf. EEA 2003, p. 14). Effectively monitored over time, the indicators could provide fruitful information about the improvement or degradation of the represented phenomenon.

In order to determine the impact on climate change, the carbon footprint (gCO_2eq/kWh) of a specific PV installation option is used. It is derived from the LCIA midpoint indicator "IPCC 2007 GWP" where GWP abbreviates global warming potential. Since the EEG understands climate and environmental protection as inherently depending on each other (cf. EEG-Explanation 2009, p. 19) the second indicator, direct space requirements, will be partly connected to climate change. It sees space as an environmental good and limiting factor and measures the net GHG emissions that can theoretically be avoided per m^2 of directly occupied space in a PV system's lifetime ($NEA_{space,LT}$). The indicator therefore captures climate change from another perspective. It does not only account for the GHG emissions of the PV system itself (carbon footprint) but also for the possible emission avoidance that can be achieved by substituting conventional with renewable electricity, i. e. the higher the emission avoidance of a PV system the more efficient it is in climate protection. Moreover, incorporating the avoided GHG emissions from the average electricity mix, which is still mainly based on fossil and non-renewable fuels like coal, oil, gas or uranium, the indicator indirectly accounts for other negative environmental effects that are associated with the conventional electricity generation (e. g. sulfur dioxide, nitrous oxide, particulate matter and other emissions, resource depletion, waste storage etc.) (cf. BMU 2012b, pp. 14-15). The relation of the $NEA_{space,LT}$ to the direct space occupation of the PV system is another efficiency component. It is based on the idea that the area of high yield locations for PV is limited and will be exploited by investors primarily because of higher returns on investment. Consequently, if the FIT PV support were oriented on space efficient criteria, more capacity could be installed at high yield locations leading to an increased electricity generation from PV or, alternatively, to the same amount of electricity with less material requirements/environmental pressures. Finally, to represent resource conservation the EPBT was chosen which based on the LCIA indicator "cumulative energy demand" (CED).

Tab. 3.1 summarizes the EEG goals considered and the criteria and indicators derived from these with their specific unit of measurement.

Tab. 3.1: Goals, criteria and indicators considered in the analysis; Source:
Authors

	No. 1	No. 2	No. 3
§1 EEG Goals	Climate protection	Environmental protection	Fossil resource conservation
Criterion	Climate change	Avoided climate change/Space	Primary energy consumption
LCIA midpoint indicator	GWP IPCC 2007 100 years v. 1.0.2 ($kgCO_2$eq)	GWP IPCC 2007 100 years v. 1.0.2 ($kgCO_2$eq)	Cumulative energy demand v. 1.0.8 (MJ_{prim})
Indicator for analysis	Carbon footprint (gCO2eq/kWh)	$NEA_{space,LT}$ (kgCO2eq/m²)	EPBT (years)

Since socio-economic interfaces with the environment are complex there
are multiple indicators to describe them (cf. EEA 1999; 2003). In PV LCA
literature carbon footprint and EPBT are the most often used indicators
to characterize PV product systems (cf. Peng et al. 2013; Turney and
Fthenakis 2011; Laleman et al. 2011). However, Turney and Fthenakis
(2011) additionally identify hazardous material emissions, land use intensity, water usage and wildlife impacts. Beylot et al. (2012) for instance use endpoint impact indicators to characterize large scale PV systems according to climate change, human health, ecosystem quality and
resources. Alsema et al. (2006) found the EPBT, GHG emission mitigation, toxic emissions, resource supply and health and safety risks to be
relevant when discussing the environmental impact of PV technologies.
The multitude of approaches to assess environmental impacts by carrying out an LCA highlights the importance of choosing the right indicator
in the right case. Therefore LCA guidelines and standards generally re-
quest to link LCA methodology and impact assessment methods to the
goal of the study (cf. ECJRC 2010, p. 29, 33; EN ISO 14044:2006, p. 20). I.
e., when intending to assess the EEG remuneration scheme according to
environmental efficiency, the LCA, on which the indicator development
is based, should be specifically designed to meet this intention.

Niemeijer and de Groot (2008) argue that *"which indicators are considered
highly influences conclusions as to whether environmental problems are serious
or not, whether conditions are improving or degrading, and in which direction*

causes and solutions need to be sought." (p. 19) McCool and Stankey (2004) state that *"because the choice of goals is a political and prescriptive action, the selection of appropriate criteria and indicators [...] is at once both a technical and normative decision."* (p. 297) Therefore, the choice of a suitable indicator and base for comparison should comprise requirements from the political and technical sphere.

According to United States National Research Council (NRC 2000) cited by the Millennium Ecosystem Assessment (MEA 2005, p. 50) an effective indicator should, among others, be able to answer the following questions:

a)   Does the indicator provide information about changes in important processes?

b)   Can the indicator detect changes at the appropriate temporal and spatial scale without being overwhelmed by variability?

c)   Is the indicator based on well-understood and generally accepted conceptual models of the system to which it is applied?

d)   Are reliable data available to assess trends and is data collection a relatively straightforward process?

e)   Are monitoring systems in place for the underlying data needed to calculate the indicator?

f)   Can policymakers easily understand the indicator?

These criteria comprise both, technical issues of measurability, temporal and spatial applicability, theoretical foundation, data availability, the possibility of monitoring but also representativeness or validity and political usability/relevance of the indicator. For the indicators exemplary used in this study, their potential relevance (Question 1) has been already discussed above. The issue of integrity will be addressed below in more detail. Their robustness (Question 2) is assessed in a sensitivity analysis (see Section 3.4) where different assumptions in the LCA modeling scheme are reconsidered. The results derived from the LCA study are based on a well established methodology incorporated in the EcoInvent 2.2 database (cf. Frischknecht et al. 2007) with which the PV systems under consideration were modeled and analyzed (Question 3). Since EPBT and GWP are most commonly used in LCA literature, a good theoretical foundation can be presumed. Finally, questions five and six, namely how the proposed indicators could be monitored and if they can be useful for a communication to policymakers will be addressed in the discussion (Chapter 5).

Besides the technical issues of "what can be measured" by an indicator (Section 3.1), the importance of the analysis' goal and thus the question of "what should be measured" by indicators (cf. McCool and Stankey 2004) has to be considered as well. McCool and Stankey (2004), discussing the derivation of meaningful sustainability indicators, underpin the importance of determining adequate goals in order to be able to identify relevant and valid indicators. These aspects seem obvious at a first glance but entail major problems in case of non-existence. It becomes apparent when reconsidering the aim of environmental protection outlined by the EEG (compare Tab. 3.1) and the criterion and indicator chosen to represented it in this study (space requirements and $NEA_{space,LT}$). The broad definition of the goal leaves room for choosing appropriate indicators. As stated by Rametsteiner et al. (2011) *"decisions on indicators are taken by a usually rather limited number of persons, which are often experts in a specific area. These experts decide on the relative importance of an issue compared to a wide range of issues available [...]"* (p. 62) Therefore the indicator selection or development process would be an essentially normative decision since it depends on the actors involved, their knowledge, intentions and perceptions (cf. Rametsteiner et al. 2011).

Here space and climate are chosen as representatives of environmental protection for a number of reasons: (1) space requirements are already topic of the PV debate in the EEG, but so far only concern space consumption in combination with land use instead of an analysis of efficient space use to reach the capacity expansion goals (cf. Reichmuth 2011 and Staiß et al. 2007), (2) climate change and natural protection are inherently linked in the understanding of the legislator, indicating that improved climate change mitigation would also improve the state of the environment (cf. EEG-Explanation 2009, p. 19) and (3) to highlight the importance of indicator selection or determination of a reference unit in indicator measurement (e. g. values per m^2 instead of per kWh) on the comparison results and consequently possible policy implications (Chapter 5). Nevertheless, it remains questionable if the $NEA_{space,LT}$ can be considered a valid indicator to measure environmental protection since it covers only a fraction of what is considered environmental problem. Therefore, when aiming to measure the efficiency of environmental protection induced by the EEG a broader designed indicator or a set of indicators, possibly accounting for the impacts at the end- instead of the midpoint (damage related indicator) might be more appropriate. Since the selection still requires prioritization and choice of certain environmental impacts, the assessment could be based on the elaborations of United Nations Environment Programme (2010, p. 37) where emissions of GHG, eutrophying substances and toxic substances, the extraction of abiotic and biotic resources as well as the use of land and fresh water are derived as the most important pressures on the environment exerted by human activities. With the current state of LCA methodology and data availability these criteria could be mirrored in suitable indicators.[1] For reasons of simplicity such an exhaustive analysis, which would also have covered the EEG goals concerning climate change and resource consumption, is not carried out here but seems fruitful in case of an advancement of ideas presented in this study. Moreover, the focus of analysis could certainly be extended from the environmental to a sustainability oriented point of view, also accounting for social and economic impacts taking into account new developments in LCA assessment methodology, namely social LCA and life cycle costing (cf. UNEP 2009; Swarr et al. 2011).

3.2 Vertices of the conducted LCA Study

LCA, as comprehended in the EN ISO 14040:2006 and EN ISO 14044:2006 standards, has been used as a methodological framework

---

[1] For a compilation of current LCA software and database for environmental impact modeling see Töpfer (2012, pp. 18-21).

for assessing the potential environmental impacts of eight different PV system installation alternatives using Mono-Si, Multi-Si and CIGS PV modules manufactured at Hanwha Q CELLS and Hanergy Solibro respectively.

In order to obtain these results a comprehensive LCI on Mono and Multi-Si modules, CIGS thin-film laminates and the corresponding supply chain was built upon producer data of Hanhwa Q CELLS, Solibro and their business partners. The considered PV module technologies are the ones currently dominating the PV market. According to EPIA (2011 p. 26) c-Si technologies, comprising Mono- and Multi-Si as well as crystalline ribbon modules have a market share of approximately 80% (date 2010).[2] The contribution of CIGS thin-film is currently still marginal but expected to grow to approximately 13% in 2020 (cf. EPIA 2011, p. 26). Fig. 3.1 shows a simplified flow chart of the life cycle of c-Si PV modules (left side) and CIGS modules (right side) and the corresponding reference flows.[3] The functional unit,[4] i. e. the reference measure of the service provided by the PV systems is the amount of *kWh* fed into the electricity grid. It can therefore be considered as a base of comparison between the different product systems studied. Process steps in the life cycle for which self raised producer data was included in the LCI model are marked with a frame. Those processes are termed foreground processes (cf. IEAPVPS 2011). All other up- and downstream activities included are accordingly termed background processes (cf. IEAPVPS 2011) and were utilized from the EcoInvent 2.2 LCI database mainly elaborated by Jungbluth et al. (2010). The attributional[5] LCI modeling was carried out with the SimaPro 7.3 LCA software. For the sum-

---

[2] Silicon ribbon PV modules were not part of the LCA. Their market share, as a part of c-Si technologies, amounted to 5% in 2010 (cf. EPIA 2011, p. 26).

[3] Reference flow is defined in LCA nomenclature as the *"measure of the outputs from processes in a given product system required to fulfill the function expressed by the functional unit used in the LCI model"* (EN ISO 14040:2006).

[4] Functional unit is defined as the *"quantified performance of a product system for use as a reference unit"* (EN ISO 14040:2006).

[5] *"Attributional modeling makes use of historical, fact-based, measurable data of known (or at least knowable) uncertainty, and includes all the processes that are identified to relevantly contribute to the system being studied."* (ECJRC 2010a, p. 71)

marizing purpose of this chapter the life cycle stages are only briefly addressed in the following.

Quartz sand is the basic material used for fabrication of semiconductor silicon (see Fig. 3.1). In a first step the sand is processed to metallurgical silicon (MG-Silicon). Afterwards it is purified in electronic, solar and off-grade silicon, which differ in their degree of purity. For the application in c-Si PV cell production a mixture of these silicon feedstocks is used. In a next step Mono- and Multi-Si ingots are grown and afterwards sawn into the well known square shaped wafers which are the base material for PV cell production. Differing crystallization processes and also feedstock material distinguish Mono-Si from Multi-Si wafers (cf. e. g. Ferazza 2012, p. 87-89). Moreover, Mono-Si wafers are more energy intensive to produce but yield better cell efficiencies due to the fact that they are perfect crystals and count with less impurities than lower quality Multi-Si cells (cf. Tobías et al. 2003, p. 283).

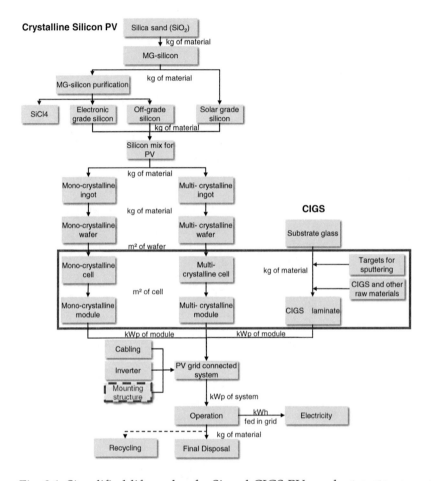

Fig. 3.1: Simplified life cycle of c-Si and CIGS PV product systems considered in the LCI model; Source: Authors

Wafers are the input material of the cell production process for which data from Hanwha Q Cells' German plant was available. Fig. A.1 in the Annex redraws the cell and module production steps of c-Si modules in a more detailed process flow chart. Wafers are treated in several steps containing chemical baths for surface etching, isolation, phosphorous diffusion, deposition of an anti-reflection-coating which gives the PV cell the typical blue or black color, contact sieve printing and firing. The finished cell is packed and shipped to module producers or "converters" which are, in the case of Hanwha Q CELLS, located in Europe and Asia. At the converter 60 cells are interconnected, embedded between different plastic sheets and a glass pane, aluminum framed and fitted with a junction box (JBox) that allows to connect the modules with each other. Such a complete PV module can be shipped to the installation site.

In case of CIGS the main life cycle route (see Fig. 3.1) is shorter than the one of c-Si PV modules because almost the whole assembly of raw and complementary input materials to a thin-film module takes place at one manufacturer. Basic materials for production are two glass panes that encompass the semi-conductor material. On one glass pane, the substrate glass, thin layers of molybdenum, CIGS, cadmium sulfide and zinc oxide are applied in different sputtering and structuring processes. The glass edges are then blasted with sand and corundum to ensure proper isolation and removal of unwanted thin-film layer residues. In a marriage process substrate and cover glass as well as a sheet of ethylvinylacetate foil are attached to each other and laminated. Finally a JBox is fitted on the module's backside and the products are packed for shipping to the installation site.

The module products of Hanwha Q CELLS and Solibro that have been assessed are displayed in Tab. 3.2. C-Si modules have a size of $1.67m^2$ while CIGS modules are addressed in a $0.94m^2$ and a smaller $0.75m^2$ version. The reference power classes, i. e. the maximum electric output of the PV module, were set to $255Wp$ in case of the Mono-Si Q.Peak, $245Wp$ for Multi-Si Q.Pro, $105Wp$ for Q.Smart L and $85Wp$ for Q.Smart UF. Q.Peak is produced at a European module converter of Hanwha Q CELLS and uses Q6LMXP3 cells with a thickness of $200\mu m$. Q.Pro modules in turn are manufactured at the European (13%) and an Asian converter (87%). They deploy either Q6LPT3 or Q6LTT3 cells. However, in the LCA an "average module" for Q.Pro is considered based on Hanwha Q CELLS production data, incorporating a mixture of both converters, cell types and cell thicknesses of 180 and $200\mu m$. The module products are named for the further analysis according to their cell technology, i. e. Mono-Si, Multi-Si, CIGS L and CIGS S, where "L" stands for "large" and "S" for "small".[6]

Tab. 3.2: Module and cell products of Hanwha Q CELLS and Solibro considered in the LCA; Source: Authors, partly accessible in Hanwha Q CELLS (n.d.) and Solibro (n.d.)

	Q.Peak	Q.Pro	Q.Smart L	Q.Smart UF
PV Technology	Mono-Si	Multi-Si	CIGS	CIGS
Type of cells (see below)	Q6LMXP3	Q6LPT3 and Q6LTT3	-	-
Amount of cells	60	60	-	-
Size (mm) (Height x Width x Depth)	1,670x1,000x50	1,670x1,000x50	1,190x790x7.3	1,190x630x7.3
Total size (m²)	1.67	1.67	0.94	0.75
Effective size (m²)	1.46	1.46	0.94	0.75
Weight (kg)	20	20	16.5	13.2
Nominal power ranges (Wp)	245-265	230-245	95-115	75-95
Reference nominal power (Wp)	255	245	105	85
Efficiency ranges	14.7%-15.9%	13.8%-15%	10.1%-12.2%	10%-12.7%
Reference efficiency	15.3%	14.7%	11.2%	11.3%
Amount of modules per kWp	3.92	4.08	9.52	11.76
Area coverage per kWp (m²)	6.55	6.82	8.95	8.82

	Q6LMXP3	Q6LPT3-G2	Q6LTT3-G2
Wafer type	Monocrystalline silicon	Multicrystalline silicon	Multicrystalline silicon
Size (mm) (Length x Width)	156x156	156x156	156x156
Area (m²)	0.024336	0.024336	0.024336
Thickness (µm)	200	160-200	160-200
Nominal power ranges (Wp)	4.14-4.58	3.75-4.23	3.75-4.23
Efficiency ranges	17%-18.8%	15.4%-17.4%	15.4%-17.4%
Frontside	3 galvanic busbars, alcaline texturization	3 busbars, acid texturization	3 busbars, acid texturization
Backside	3x6 solder contacts (silver, aluminium), aluminium back contact (pad cell)	3x6 solder contacts (silver, aluminium), aluminium back contact (pad cell)	3 run-through bus bars

---

[6] At the time of data collection and LCA elaborations (2010-2012) Solibro has been a subsidiary of the former Q-Cells SE, which is now integrated in the Hanwha Group (Hanwha Q CELLS). Since September 2012 Solibro GmbH, in turn, is a part of the Hanergy Holding Group located in China. After the spin-off the CIGS module products "Q.Smart UF" and "Q.Smart L" have been renamed to "SL 1" and "SL 2" respectively.

The modules presented in Tab. 3.2 can be installed and connected to the grid in diverse ways. In the LCA a set of eight installation types is considered. They range from mounting systems typically found on residential rooftops, in or on facades, on commercial and industrial flat roofs and in utility scale open ground PV parks. The dashed frame in Fig. 3.1 indicates that the studied reference mounting systems stem either from established datasets of the EcoInvent 2.2 database (cf. Jungbluth et al. 2010) or system installations for which data could be obtained from Hanwha Q CELLS and Solibro. Specifications of the systems addressed and corresponding module alignment, tilt, mounting technique and corresponding data source are shown in Tab. A.1 in the Annex.

Residential roof solutions are represented by mounted and integrated slanted roof PV systems, optimally aligned towards the sun. Moreover three flat roof mounting solutions are studied, being either mounted in rows with optimal tilt towards the sun, in a tent-like structure using two modules facing east- and westwards or mounted horizontally on a flat roof. In the latter alternative the system's inclination towards the sun is precisely not 0° but 3° to ensure proper rain and snow runoff. The assessed facade systems contain a version which is mounted *on* the building's facade and one which is integrated *in* the facade. All mentioned varieties have been studied in options equipped with Mono- and Multi-Si as well as CIGS PV modules. The open ground systems under assessment represent an exemption because of a difference in thin-film and c-Si module installation. Both types are optimally oriented towards the sun but in the thin-film case four modules are mounted on top of each other. In the c-Si case only two modules are installed on top of each other as it is depicted in Tab. A.1. Material input data of the mounting structures for "flat roof (east/west)" and "flat roof (horizontal)" systems was raised from installations at the premises of Hanwha Q CELLS and Solibro respectively. The systems are assumed to be linearly scalable, indicating that the actual size of the PV system under consideration is roughly proportional to the potential environmental impacts derived below (Section 3.3).

PV systems basically consist of PV modules, collecting the solar energy and directly converting it into electricity, and of balance of system (BoS) components, which assure that the electricity is generated efficiently and can be utilized. BoS parts comprise cabling and mounting structure as well as an inverter, converting the direct current delivered by the PV modules into average current that can be fed into the electricity grid, or batteries which store the produced electricity in stand-alone systems (cf. e. g. Preiser 2003, pp. 784-790). Life cycle data on inverters and cabling was obtained from the EcoInvent 2.2 database.

Foreground processes in the LCI include the life cycle of raw materials for production, packaging, transport activities, waste and production halls among others. Additionally the energy and media inputs (e. g. heat, electricity and water) and emissions in different processes are accounted for. The supply chain from wafer to readily installed module was modeled according to Hanwha Q CELLS' global business relationships and module sales in the reference year 2011. The same applies to CIGS PV systems except from the reference year which was 2010.

The use phase in which the PV system feeds electricity into the grid was assumed to be 30 years for modules, mounting structure and cabling, and 15 years for the inverter (cf. IEAPVPS 2011). The reference location to which the LCA results apply is Munich. Underlying irradiation and insolation values and climate data stem from the software "PVSol Expert 4.5 (R1)" which was used for PV yield simulations. The climate data considered is based on observations in Munich from 1981 until 2000. Yearly degradation (or power loss) of the PV modules was set to 0.6% according to Hanwha Q CELLS experiences. A complete documentation of the required reporting parameters of the International Energy Agency Photovoltaic Power System Programme (IEAPVPS 2011) is provided in Tab. A.2 of in the Annex. Additionally Tab. A.3 in the Annex summarizes the $ASAY$ of the considered PV installation/technology combinations at the reference location. It represents the averaged electricity yield over the modules' lifetime taking the assumed degradation rate into account.

At the end of life (EoL) of the PV systems a complete final disposal was assumed. Recycling processes could not be included due to recently lacking LCI data (cf. Alsema 2012, p. 1099; Berger et al. 2010; Jungbluth et al. 2010, p. 68). The system boundaries of the LCI illustratively summarize what has been included in the LCA and what has been left out. They are shown in Fig. A.2 and Fig. A.3 in the Annex for c-Si and CIGS product systems respectively.

3.3 Update of Indicator Results

Due to the iterative character of LCA (cf. ECJRC 2010a, p. 7) the results obtained in Töpfer (2012) were updated with more recent data becoming available after finishing the previous study. Moreover the LCA has been critically reviewed internally at Hanwha Q CELLS and by the authors resulting in the recalculation of some process in- and outputs in the LCI. Below the updated results for the three decisive indicators, carbon footprint, $NEA_{space,LT}$ and EPBT are outlined. The EPBT is assessed first in order to be consistent with the work in Töpfer (2012) and for reasons of interconnectedness with the carbon footprint. The subsequent sensitivity analysis (Section 3.4), literature review (Section 3.5) and relation to the FIT's of the EEG (Chapter 4) will be based on the LCA outcome presented here.

3.3.1 Energy Payback Time

In order to avoid a misinterpretation of results it is worthwhile to briefly address the composition and influential factors of the EPBT. Based on the definition of Alsema (2012, p. 1099) the EPBT is the ratio of the gross primary energy requirement of a product system divided by its annual primary energy output. Therefore, it represents the time period a PV system needs to recover the energy input during its life cycle. The nominator of the EPBT equation can therefore be considered as the compilation of the CED of all input materials (E_{mat}) plus manufacturing

(E_{manuf}), transport (E_{trans}), installation (E_{inst}) and end of life (E_{EoL}) processes. Whereas the denominator consists of the "generated" average net end energy, or $ASAY$ (to be found in Tab. A.3 in the Annex) times a conversion factor that transforms end to primary energy R_{prim}. Thus the EPBT can be represented by Formula (3.1) (adapted from Fthenakis and Kim 2010 and Nishimura et al. 2010):

$$EPBT = \frac{E_{mat} + E_{manuf} + E_{trans} + E_{inst} + E_{EoL}}{(ASAY) * R_{prim}} \qquad (3.1)$$

For the reference location of Munich R_{prim} amounts to 11.6 $\frac{MJ_{prim}}{kWh_{end}}$. Put differently the product in the denominator can be interpreted as the yearly avoided electricity in terms of MJ primary energy from the conventional German electricity mix.

The CED required as an input in the EPBT calculation was obtained from the LCIA, i. e. it is influenced by the LCA's scope, assumptions and methodology. The annual energy output in turn is based on a yield simulation of the specific system assessed and installed at the reference location. It is mainly influenced by the solar irradiation impinging on a defined area of the PV system. The amount of this "insolation" is a function of latitude, air mass, incidence angle and diffusion of solar radiation in the PV module (cf. Pacca et al. 2007). Furthermore, module efficiency and transmission losses play a remarkable role (cf. Jahn and Nasse 2004). Since the systems under consideration differ in their inclination angle towards the sun (see Tab. A.1) and different module technologies are considered within the installation options (see Tab. 3.2) the resulting insolation and electricity outputs diverge.

Tab. 3.3 represents the updated outcome for specific system/module combinations installed in Munich. EPBTs range from 1.43 years in the best case (green) to 3.67 years in the worst case (red). It can be seen that generally CIGS modules perform best with an energetic amortization which is approximately one year shorter than for c-Si modules. Due to higher primary energy requirements the Mono-Si installations lag behind those having installed Multi-Si modules despite their high electric

Tab. 3.3: EPBT results of considered PV systems and technologies; Source: Authors

EPBT (years)	Multi-Si	Mono-Si	CIGS L	CIGS S
Slanted Roof (mounted)	2.31	2.43	1.46	1.46
Slanted Roof (integrated)	2.38	2.51	1.48	1.46
Flat Roof (normal)	2.26	2.37	1.46	1.43
Flat Roof (east/west)	2.62	2.76	1.64	1.61
Flat Roof (horizontal)	2.38	2.51	1.52	1.49
Facade (mounted)	3.32	3.49	2.10	2.09
Facade (integrated)	3.49	3.67	2.23	2.22
Open Ground	2.40	2.49	1.70	1.68

efficiencies. When comparing installation systems among each other, facade solution can be identified as the worst performers due to the unfavorable module tilt of 90° resulting in lower insolation on the PV array and thus lower electricity generation. Flat roof (normal) systems show the best results followed by slanted roof (mounted and integrated) and the horizontally installed flat roof version. Despite its high material inputs the open ground system's EPBTs are not significantly higher as those found for residential or industrial scale building mounted systems, being attributable to the high yield in the operation phase. The tent-like structure of flat roof (east/west), with module tilts of 10° in both directions, is the reason for its rather weak performance in terms of EPBT since the optimal module tilt towards the sun would have been 35°.

To gain more insight in the composition of the EPBT Fig. 3.2 exemplary depicts the contribution of different system parts to the total of a slanted roof (mounted) system. They are almost equally distributed when comparing Mono-Si to Multi-Si and CIGS L to CIGS S.

In the latter case only small divergences in the processing at Solibro and the amount of glass, structure and cabling needed can be found. These are explained by size differences and their production in different halls using slightly varying inputs. For c-Si modules a higher deviation between the slanted roof system equipped with Multi-Si modules in comparison to the one equipped with Mono-Si modules can be identified. The high energy requirements of the Mono-Si wafer are the explanation

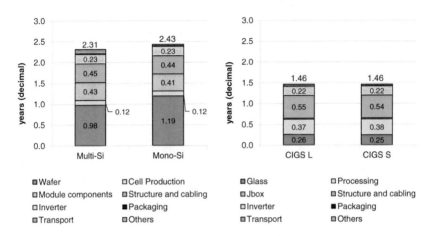

Fig. 3.2: Contribution of different system parts to the EPBT performance of a residential slanted roof (mounted) system; Source: Authors

for the low performance of the total system; even offsetting the lower impact resulting from less required installation material, cabling, transport efforts and packaging. Additionally the higher electricity yield of the Mono-Si modules cannot compensate the high CED in the production phase.

3.3.2 Carbon Footprint

After having analyzed the EPBT of all considered PV systems, the carbon footprint indicator results are addressed in the following. Formula (3.2) is the basis for its calculation:

$$CarbonFootprint = \frac{LCE}{ASAY * LT} \qquad (3.2)$$

where LCE represents the life cycle CO_2eq emissions which are divided by the product of $ASAY$ (Tab. A.3) and the lifetime (LT) of the system (30 years). LCE are obtained from the LCA using the "IPCC 2007 GWP, 100 years" LCIA method in the EcoInvent 2.2 database (cf. Hischier et al. 2010, pp. 136-142).

The updated results can be seen in Tab. 3.4. When comparing the colors to those of the EPBT above basically the same distribution can be identified, indicating that EPBTs and carbon footprints are closely correlated. Facade solutions perform worst, followed by the flat roof (east/west) system and the open ground system. Slanted roof and remaining flat roof installations can attain the best carbon footprints. CIGS modules emit in general approximately $20gCO_2eq/kWh$ less than their c-Si counterparts. Differences between both CIGS modules can be considered as marginal. Whereas the differences between Multi- and Mono-Si are more salient, with the former achieving approximately $3\text{-}5gCO_2eq/kWh$ lower emissions than the latter. Concerning absolute emissions values it can be found that those systems equipped with CIGS modules range from 30 to $45gCO_2eq/kWh$ and those with c-Si modules from 47 to $78gCO_2eq/kWh$.

Since both considered PV manufacturers deploy electricity from renewable energies for their production processes in Germany, the outlined carbon footprints are reduced significantly in the corresponding life cycle phases. This is especially the case for Solibro where the main material and energy intensive production steps take place in Germany. Conversely the cell production process taking place at Hanwha Q CELLS Germany is only marginally contributing to the total impact and therefore does not significantly improve the c-Si system's performance. These aspects are displayed in detail in Fig. 3.3 for an exemplary slanted roof

Tab. 3.4: Carbon footprint indicator results of considered PV systems and technologies; Source: Authors

Carbon Footprint (gCO$_2$eq/kWh)	Multi-Si	Mono-Si	CIGS L	CIGS S
Slanted Roof (mounted)	48.11	51.39	30.08	29.03
Slanted Roof (integrated)	49.41	52.92	30.13	29.20
Flat Roof (normal)	47.23	50.36	29.92	28.88
Flat Roof (east/west)	53.61	57.56	32.26	31.04
Flat Roof (horizontal)	49.11	52.57	30.37	29.24
Facade (mounted)	69.03	73.82	43.09	41.50
Facade (integrated)	72.75	77.73	45.91	44.24
Open Ground	51.57	54.11	36.98	35.94

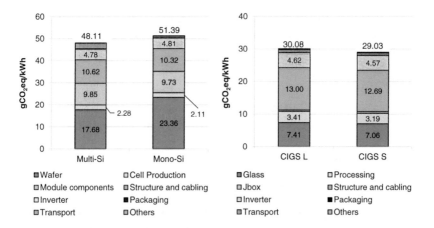

Fig. 3.3: Contribution of different system parts to the carbon footprint
performance of a residential slanted roof (mounted) system;
Source: Authors

(mounted) system.[7] It can be seen that the energy intensive wafer production is the reason why the Mono-Si system shows worse indicator
results than the Multi-Si option, with the remaining contributions being similar. In total the Mono-Si slanted roof system would emit 51.4
gCO_2eq per produced kWh, while the Multi-Si slanted roof system emits
approximately 3 gCO_2eq/kWh less. The slanted roof systems utilizing
CIGS modules can achieve additional GHG emission savings of approximately 20 gCO_2eq/kWh. The mounting and cabling structure as well as
the glass panes contribute significantly to the carbon footprint, the first
with approximately 43% and the latter with 25%. The influence of the
processing can be reduced to about 3 gCO_2eq/kWh or 11% by using renewable electricity, which is 13%-points lower than in the EPBT results.
Differences between the CIGS options are only marginal.

---

[7] When comparing the four bars shown in Fig. 3.3 the reader is encouraged to notice the different
scales on the ordinate.

3.3.3 Space Requirements

The last indicator under assessment is the $NEA_{space,LT}$ measuring the net lifetime GHG emission avoidance per m^2 space requirement[8] of the specific PV technology/installation combination seeing space as an environmental good and limiting factor. Leaving out constraining factors can be misleading since some PV systems are particularly designed to maximize their yield while minimizing the occupied area. This is e. g. the case for rooftop systems where space is an important factor in the system design. For this reason the considered PV system installations were assessed according to their space intensity. The resulting space requirements per kWp installed system are summarized in Tab. A.4 in the Annex. Differences occur because of the type of module technology chosen. Taking a $3\,kWp$ residential slanted roof (mounted) system with Mono-Si and CIGS S modules as an example, approximately 12 Mono-Si modules or 36 CIGS S modules would be required to install the same system capacity, resulting in a respective roof space consumption of 20 m^2 or 27 m^2.[9] Assuming that a house owner in Munich has an available roof space of $20m^2$ she can expect to harvest around 3,228 $kWh/year$ choosing Mono-Si modules and 2,545 $kWh/year$ with CIGS S modules (compare to Tab. A.5 in the Annex). Since the annual yield is not only an influential parameter of the environmental system performance (e. g. the carbon footprint) but also its economic amortization, it is important to consider the available space as a limiting factor for PV systems. This becomes even more remarkable when considering different system designs. A flat roof (normal) system or open ground system for example, are built in rows lined up consecutively and are tilted 35° towards the sun. This results in necessary shading spaces between the module rows which, in turn, enlarge the system's space occupation. However, there are designs like the flat roof (east/west) installation specifically creat-

---

[8] Space requirement is defined as the effective area that is used up by the PV system. In case of rooftop installation this area refers to the amount of roof, in case of facade solutions the amount of the building's facade and in case of open ground PV parks the amount of land area that is occupied.

[9] Example based on the assumptions made in Tab. 3.2.

ed to dismiss space losses through shadings, accepting an unfavorable module inclination of 10°.

In order to calculate the third indicator the following formulas were applied:

$$NEA_{space,LT} = GEA_{LT} - LCE \qquad (3.3)$$

where $NEA_{space,LT}$ is the net emission avoidance and $GEA_{space,LT}$ the gross emission avoidance in a PV system's lifetime and LCE again the CO_2eq emission in a PV system's life cycle. All these are expressed in $kgCO_2eq$ per m^2 occupied space. The $GEA_{space,LT}$ in turn was calculated with the formula:

$$GEA_{space,LT} = \frac{ASAY * EF * LT}{SR} \qquad (3.4)$$

where $ASAY$ represents the annual electricity yield per kWp of system (Tab. A.3 in the Annex), SR the space requirement for 1 kWp of system (Tab. A.4 in the Annex), EF the emission factor of the German average electricity mix[10] and LT the system's lifetime of 30 years. Resulting values for $NEA_{space,LT}$ can be found in Tab. 3.5.

Tab. 3.5: $NEA_{space,LT}$ indicator results of considered PV systems and technologies; Source: Authors

$NEA_{space,LT}$ in $kgCO_2eq/m^2$	Multi-Si	Mono-Si	CIGS L	CIGS S
Slanted Roof (mounted)	2,606.03	2,685.04	2,103.48	2,173.14
Slanted Roof (integrated)	2,494.98	2,566.56	2,038.15	2,105.92
Flat Roof (normal)	1,058.79	1,092.31	852.44	873.62
Flat Roof (east/west)	2,225.31	2,288.47	1,801.41	1,863.30
Flat Roof (horizontal)	2,464.28	2,538.07	1,977.95	2,045.06
Facade (mounted)	1,733.52	1,781.51	1,404.72	1,454.68
Facade (integrated)	1,664.06	1,707.70	1,360.88	1,409.19
Open Ground	1,077.06	1,115.31	876.72	896.36

---

[10] The emission factor used is 657 gCO_2eq/kWh obtained from the EcoInvent 2.2 database with the GWP, IPCC 2007 100a LCIA method.

As it could already be suspected in the above-mentioned example, the effect of including a restraining factor in the analysis is responsible for a significant change in the results compared to the EPBT and carbon footprint. When comparing module technologies among each other it can be seen that the CIGS panels perform much worse than c-Si modules. The Mono-Si panel, counting with the worst results in the EPBT and carbon footprint, is the best available option according to the $NEA_{space,LT}$. Mutli-Si modules achieve only marginally less $NEA_{space,LT}$ and thereby outperform CIGS modules as well. The explanation for these results is based on the lower module power/module area ratio: CIGS modules achieve a ratio of 111-113 Wp/m^2 while c-Si panels attain ratios from 146-152Wp/m^2.

When having a closer look at the different installation options, the flat roof (normal) and open ground systems show the lowest emission saving/space ratio. Every m^2 of space used can in these cases avoid approximately 1 tCO_2eq when utilizing c-Si PV panels and around 900$kgCO_eq$ using CIGS modules. In the best cases the $NEA_{space,LT}$ of PV systems can be increased up to 2.2 to 2.7 tCO_2eq. For flat roofs the most preferable option would therefore not be to install the system in a conventional way, i. e. lining up 35° inclined modules behind each other, but using the horizontal or east/west aligned system design.

3.4 Sensitivity Analysis

The developed LCA model was based on a set of assumptions with high potential influence on the indicator results presented above (Section 3.3). Therefore, this section is dedicated to critically scrutinize some of the assumptions made in the LCA model, namely, varying the c-Si wafer electricity consumption and thickness (Section 3.4.1), repealing the assumption that building integrated PV (BIPV)

Some assumptions of the LCA outlined in Töpfer (2012, pp. 28-29) will not be addressed here since they were found to have no significant contribution to the indicator results or because no reliable data on these processes was available. The former refers in particular to transport distances, the assumption that c-Si building integrated PV modules are integrated with an aluminium frame and the type of EoL disposal, whereas the latter mainly addresses the no-recycling assumption for which currently no workable LCI data exists (cf. e. g. Berger et al. 2010 or Jungbluth et al. 2010, p. 68). This is mainly attributable to the recently low amounts of PV waste (cf. McDonald and Pearce 2010) but will be of high interest in the future; not only because of environmental and social sustainability issues but also due to the increased resource consumption and related costs for the PV industry (cf. Zuser and Rechberger 2011).

3.4.1 Crystalline Silicon Wafers

From the results outlined above it became apparent that the wafer of c-Si modules is the product part with the highest contribution to the environmental impact (cf. Fig. 3.2 and Fig. 3.3). Hence, it has to be scrutinized how a variation in the wafer production and input in the c-Si cell production can influence the indicator results. Since the wafer inventories were used as background processes in the LCI model they were based on assumptions made by the authors of these inventories and the time when their data was collected. Because of the fast growth of the PV industry in the last decade and subsequent capacity enlargements in silicon, wafer, cell and modules production (cf. Jäger-Waldau 2012), the corresponding data sets in LCA databases used[11] might already be outdated.

In case of Mono-Si ingotting the EcoInvent 2.2 data set accounts for an electricity consumption of the Czochralski (CZ)-ingot growth of 200 kWh/kg for electronic grade silicon and 85 kWh/kg for PV grade silicon (cf. Junbluth et al. 2010, pp. 40-41). A more recent publication by Ferazza (2012, p. 88) states that producers already reported figures of 40 kWh/kg and that 100 kWh/kg could be seen as a rule of thumb

---

[11] The LCI model was based on the EcoInvent 2.2 database mainly based on the work of Junbgluth et al. (2010).

for Mono-Si ingotting. Unfortunately no verified industry data could be obtained here but at of Hanwha Q CELLS 40 kWh/kg were seen as a realistic value for the CZ-ingots grown for the PV industry. For Multi-Si crystallization processes Jungbluth et al. (2010) assume 19.3 kWh/kg electricity consumption based on literature values reported by de Wild-Scholten and Alsema (2005). At Hanwha Q CELLS 8 kWh/kg were estimated as a more recent and realistic figure.

The effect of the ingotting process adaption can be seen in Fig. 3.4 displaying the changed results for EPBT, carbon footprint and $NEA_{space,LT}$ for a residential slanted roof (mounted) system using Mono- and Multi-Si PV panels. The EPBT can be reduced by approximately 9.3% or 0.23 years for the system using Mono-Si modules and by 2.8% or 0.06 years when using Multi-Si panels, considering the assumptions mentioned above. In terms of the carbon footprint the resulting figures are similar. Reductions of 9.8% or 5 gCO_2eq/kWh and 2.9% or 1.4 $g/CO_2eq/kWh$ for Mono-Si and Multi-Si respectively are possible. Concerning the $NEA_{space,LT}$ an additional emission avoidance gain of 22 $kgCO_2eq/m^2$ can be achieved in the Mono-Si case and of 6 $kgCO_2eq/m^2$ in the Multi-Si case. Differences among the system installation types considered are negligible. They are summarized in Tab. A.6 in the Annex. The analysis reveals that an update of the background process data from the EcoInvent 2.2 database (cf. Jungbluth et al. 2010, pp. 40-46) to state of the art up-stream processes could, especially in the Mono-Si case, significantly reduce their resulting impact. Hence, the results used in this study might be an overestimation of the negative environmental impacts that occur from current PV module production.

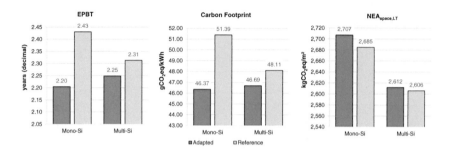

Fig. 3.4: Effects of electricity consumption in the crystallization phase on the indicator results of a slanted roof (mounted) system; Source: Authors

In addition to the electricity adaption analysis it is important to consider which wafer thickness is used in the PV cell production process. Steady cost reduction efforts by the industry lead to an endeavor of reducing the material inputs especially for expensive wafer products (cf. Szlufcik et al. 2012, p. 130). Therefore the wafer thicknesses of c-Si cells have been constantly reduced from on average 300 μm reported by Hegedus and Luque (2003, p. 24) to approximately an average of 190 μm compiled in a recent company screening by Zuser and Rechberger (2011).

At Hanwha Q CELLS the wafer thicknesses used for Multi-Si production vary from 180-200 μm depending on the supplier. In case of Mono-Si cells mainly wafers with a thickness of 200 μm were used in the reference year (2011). These cases are displayed as reference values in Fig. 3.5 (indicated by the red cross). The chart assembles the results for all three indicators depending on the wafer thickness chosen. Considered thicknesses range from 160 μm wafers, which were successfully tested in Hanwha Q CELLS production lines, to 270 μm, which is the value chosen by Jungbluth et al. (2010, p. 50) for the standard process of Mono-Si wafers in the EcoInvent 2.2 database. In case of Multi-Si cells the authors used a wafer thickness of 240 μm which is also included in this sensitivity analysis.

Fig. 3.5 shows that with increasing wafer thickness indicator results deteriorate. The EPBT measured on the lower part of the left axis can be decreased by 8% from the reference case ($200\mu m$ indicated by the red cross) when choosing thin $160\mu m$ wafers in the Mono-Si case (upper chart). Compared to what has been used by Jungbluth et al. (2010, p. 50) the EPBT of a Mono-Si system would be prolonged by 0.24 years.

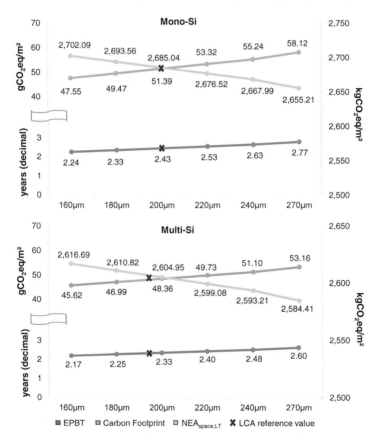

Fig. 3.5: Effects of wafer thickness adaption on the indicator results of a slanted roof (mounted) system; Source: Authors

The results are pointing in the same direction when analyzing the Multi-Si system (lower chart). Here the reference wafer mix used has an average thickness of 196 μm (indicated by the red cross) resulting in an energy payback time of 2.31 years which could be reduced by 6% when choosing only 160 μm wafers and would, in case of 240 μm wafers, be extended by 7% to 2.48 years. Carbon Footprint results show the same correlation between the different thicknesses analyzed. For the Mono-Si system they range between 47 and 58 gCO_2eq/kWh starting from the reference value of 51 gCO_2eq/kWh. The $NEA_{space,LT}$ (displayed in green) decreases by 0.3% in the Mono-Si case and 0.2% in the Multi-Si case with each step of increasing wafer thickness, except for the bigger leap between 240 μm to 270μm where the decrease is approximately 0.5% for Mono-Si and 0.3% for Multi-Si.

The carried out sensitivity analysis on different wafer thicknesses demonstrates that a variation can imply a significant change in the outcome of the EPBT and carbon footprint in particular. Therefore comparisons to other studies using a different mix of wafers in their models should be made carefully. On the other hand it might be possible to explain observable differences between different PV LCA studies with the type of wafer used. Choosing the standard processes of the EcoInvent 2.2 database (cf. Jungbluth et al. 2010, p. 53) would, from this point of analysis, lead to an overestimation of the environmental impact of PV cells produced with current technology.

3.4.2 Reconsideration of Building Integration

In the LCI, it was supposed that BIPV, i. e. the considered "Slanted roof (integrated)" and "Facade (integrated)" do not replace building parts. Since the integrated system options serve as an active part of the building structure either on the roof or in the facade this assumption will be contested by exemplary showing the effect of a substitution of conventional roof tiles with a slanted roof (integrated) PV system.

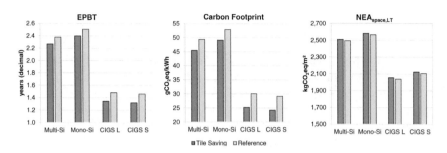

Fig. 3.6: Effects of substituting roofing tiles with an integrated PV system on the indicator results; Source: Authors

It can be argued that integrating a PV system in a building, rather than retrofitting it on the building, substitutes some of its parts, which in turn reduces the material consumption and thus environmental impact of the PV system (cf. Lu and Yang 2010). Roof integrated PV modules could for example replace clay tiles or other roof cover materials; facade solutions might substitute glass panes, bricks, plaster and paint, etc. Furthermore, Lu and Yang (2010) argue to consider the effects of thermal load changes in the building resulting from the PV system installation. Since PV systems can be integrated in a wide range of building types with diverse applications and architecture (cf. Reijenga 2003, pp. 1027-1029), the focus is set exemplary on a slanted roof PV system integration in a residential building where clay tiles are substituted.

In order to calculate the effects on the environmental PV system performance by substituting the roof tiles, company information from Nelskamp (2012) is used, assuming the roof is covered with double-troughed tiles with a weight of approximately 44.8 $kg/m^2 - roof$. The avoided environmental impacts are derived from the EcoInvent 2.2 database process "Roof tile, at plant /RER U" and subsequently deducted from the impact of 1 m^2 slanted roof (integrated) PV system. They amount to a CED of 3.9 $MJ/kg - tile$ and a GWP of 357 $gCO_2eq/kg - tile$. The effects on the indicator results, comparing the reference case to the tile saving scenario, can be found in Fig. 3.6.

The avoided life cycle environmental impacts of the tiles are responsible for a reduced EPBT of approximately 4-5% in the c-Si case and 9-10% when using a slanted roof (integrated) system with CIGS modules. Possible CO_2eq emission savings are higher and range from 7-8% for a system equipped with c-Si modules and 16-17% for CIGS modules. The influences on the $NEA_{space,LT}$ are negligible, lying all below 1%.

The analysis shows that expanding the system boundaries to avoided products can have an influence on the obtained indicator results. Since there is a huge variety of architectural possibilities for integrating solar systems in a building the effect on the environmental impact of the PV system can be that diverse as well. Therefore, when carrying out more detailed analyses on the impact avoidance of building materials by PV systems, it might be beneficial to assess specific case studies. Moreover, the sensitivity analysis again clarifies that when comparing different LCA studies a key determinant of their comparability are fairly matching system boundaries and applied methodology.[12]

3.4.3 Electricity from Renewables at Production Sites

The LCI model was built upon the production conditions of Hanwha Q CELLS and Solibro, comprising the specific mix of renewable electricity from Scandinavian hydropower stations used on the German premises. Because the results shall be used on a broader scale in this study, it seems fruitful to test the effects of substituting the specific electricity mix used with the German average electricity mix. Therefore, in all foreground processes of the LCI model which take place at Hanwha Q CELLS and Solibro in Germany the specific mix of renewable electricity is substituted by the average German electricity mix from the EcoInvent 2.2 database. The adaption implies a considerable change in some of the indicators, especially in the carbon footprint. Fig. 3.7 depicts the adapted case in comparison to the reference results exemplary for a slanted

---

[12] Transparency criteria for this LCA study according to IEAPVPS (2011) are listed in Tab. A.2 in the Annex.

roof (mounted) PV system. The deterioration of results induced by the adapted electricity mix for a PV system using c-Si modules ranges from 6-7% while the impact on CIGS modules is considerably higher, lying between 43% and 48%. This difference between c-Si and CIGS can be explained by the production conditions. While the c-Si cell production at Hanwha Q CELLS is only a marginal step in the PV modules' life cycle, almost the complete electricity intensive production of CIGS modules takes place at Solibro. The corresponding values for the EPBT and $NEA_{space,LT}$ as well as a summarizing table can be found in Fig. A.4 and Tab. A.7 in the Annex.

It can be seen that the adapted assumptions in the LCA increase the environmental impacts of CIGS PV systems relatively to those of c-Si systems so that the obtained indicator results move closer together. The magnitude of change, however, does not lead to switching results, indicating that PV systems deploying CIGS still count with a better environmental performance in case of carbon footprint and EPBT and a worse outcome when considering the $NEA_{space,LT}$.

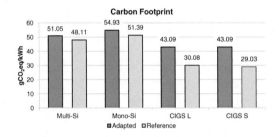

Fig. 3.7: Effects of foreground processes using the German electricity mix on the Carbon Footprint; Source: Authors

3.4.4 Lifetime and Location

Finally this section varies the assumed lifetime and installation location of the PV systems under consideration. According to IEAPVPS (2011) a

PV system's operation time amounts to approximately 30 years, which was the reference case in the LCA. Subsequently the impact of varying operation times from 5-30 years on the indicator results is plotted in Fig. 3.8 exemplary for a slanted roof (mounted) installation.

In order to obtain comparable results, the calculation for each lifetime considered refers to a electricity production time frame of 30 years. I. e. when accounting for a five year PV system lifetime, this system has to be replaced six times in 30 years, consequently increasing the CO_2eq emissions and primary energy demand by a factor of six.

With an increased PV system lifetime, decreasing EPBT, carbon footprint and increasing $NEA_{space,LT}$ can be found independently from the module technology chosen (see Fig. 3.8). All curves representing the EPBT and carbon footprint show an asymptotic behavior with CIGS modules constantly achieving better results than c-Si modules. If the PV system stopped producing electricity after 15 years of operation the EPBT of c-Si (blue dotted and golden line) and CIGS modules (dashed green line) would be almost doubled in comparison to the assumed reference lifetime of 30 years. The reaction of the carbon footprint results is even higher; a bisection of the system's lifetime entails a quadrupling of the CO_2eq emission per produced kWh. The $NEA_{space,LT}$ is more robust to changes in lifetime because of the proportionally higher influence of the avoided CO_2eq emissions in comparison to the emissions occurring from PV system production. Halving the lifetime would imply a $NEA_{space,LT}$ decrease of 3-4% in the c-Si case and 0.3% in the CIGS case. It is interesting to have a closer look at the $NEA_{space,LT}$ results deploying CIGS modules. The maximum achievable $NEA_{space,LT}$ in the reference frame of 30 years already occurs after 22 instead of 30 years, indicating on the one hand that an earlier PV system replacement could improve environmental impacts by avoiding excessive module degradation,[13] and, on the other hand, the importance of preventing fast electric degradation in PV R&D is also emphasized from an environmental point of view.

---

[13] Here an annual degradation 0.6% of the PV system's electricity yield is assumed.

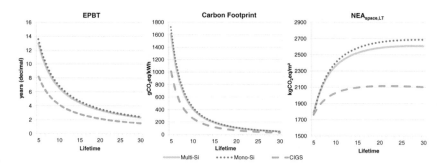

Fig. 3.8: Effects of a varied lifetime on the indicator results; Source: Authors

In a last sensitivity check the indicator results are calculated for different PV system installation locations. A change of location entails different irradiation levels to which the PV modules are exposed as well as differing electricity mixes in which the systems are embedded, ultimately leading to a diverse avoidance of conventional energy carriers and corresponding GHG emissions. The reference site being representative for the indicator results above was Munich. In order to analyze how different surroundings influence the indicators, yield data for the assessed PV module/installation combinations were obtained[14] for six additional locations, namely Cadiz (Spain), Toulouse (France), Las Vegas (USA), London (United Kingdom), Rome (Italy) and Cape Town (South Africa). Module tilting, irradiation values as well as additional location characteristics can be found in Tab. A.8 in the Annex. Fig. 3.9 depicts the EPBT, carbon footprint and $NEA_{space,LT}$ indicator results for different PV systems exemplary using Multi-Si modules.

Systems installed in Las Vegas, Cadiz and Cape Town generally perform best in terms of EPBT and carbon footprint, followed by Rome and Toulouse. Munich and especially London reveal the worst results. This sequence follows the order of local irradiation values, which are the most influential driver in the calculations. However, slight deviations can be identified in the EPBT chart and are attributable to the pri-

---

[14] PVSol Expert 4.5 (R1) Simulation.

mary energy factor (R_{prim}, see Formula (3.1)), being dependent on the electricity mix prevalent in the specific country. The higher this factor, the higher the conventional energy that can be avoided by the PV system ultimately leading to a lower EPBT. When exemplary comparing a PV system set up in Munich to an installation in Las Vegas, the latter one could reduce the EPBT by 38-47% and the carbon footprint only by 31-40%, depending on the installation type.

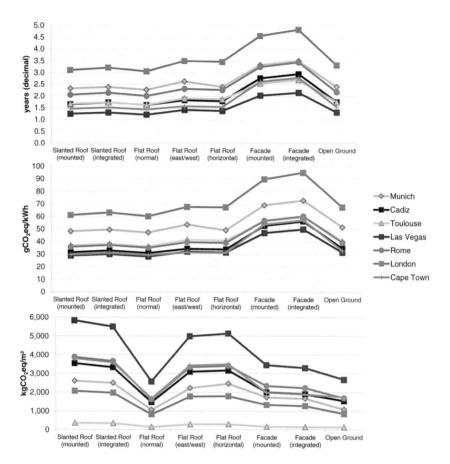

Fig. 3.9: EPBT, carbon footprint and $NEA_{space,LT}$ indicator results for PV
systems installed at different locations using Multi-Si modules;
Source: Authors

$NEA_{space,LT}$ results differ from those of EPBT and carbon footprint since
they mainly depend on the avoided emissions in the PV system's lifetime. Thus, in countries where the electricity mix count with high CO_2eq
emissions, the emissions avoidance of the PV system's electricity provision is also high and *vice versa*. This can be illustrated by the French case.

The PV system in Toulouse accounts for the lowest $NEA_{space,LT}$ because of the low emissions occurring from the French electricity mix, mainly supplied by nuclear power. Additionally, the module-yield/space-occupation ratio has an influence on the results displayed in the lower graph of Fig. 3.9. Those system types with a high yield per m^2 occupied space (e. g. slanted roof, flat roof (east/west) or flat roof (horizontal) solutions) perform, independently from the installation country, better than those systems with either an unfavorable tilt (facade solutions) or high space occupation (flat roof (normal) and open ground). Comparing the highest $NEA_{space,LT}$ values observed in Las Vegas to the ones of the reference case of Munich shows a possible indicator improvement from 98% to 148%, depending on the installation type.

To conclude, the sensitivity analysis showed that the definition of comprehensive and transparent system boundaries is of high importance for comparable and universally applicable indicators (see Fig. A.2 and Fig. A.3 in the Annex). Additionally, it became apparent that with increased technological development (e. g. higher PV system yield, thinner wafers etc.) the above-mentioned potential environmental impacts are likely to decrease and that the indicators derived are consequently not fixed but dependent on the time of evaluation in combination with a set of assumptions and uncertainties. Moreover, the carried out impact assessment contains only a limited number of environmentally significant parameters. It should therefore be noticed that the further analysis (see Chapter 4) specifically refers to the LCA results and corresponding assumptions outlined in this chapter and Tab. A.2 in the Annex.

3.5 Comparative Literature Review

In order to back up the findings of Section 3.3, they are briefly compared to those of other authors considering the same PV technologies and environmental performance criteria. In literature, plenty of investigations on the carbon footprint and EPBT of PV systems can be found; but, to the knowledge of the authors, there is only one study by Halasah et al. (2013) connecting GHG emissions with space intensity and coming up with an indicator which similar to the $NEA_{space,LT}$. Despite the remarkable amount of studies on EPBT and carbon footprint Pacca et al. (2007) argue that *"different studies use different methods, draw on different boundary conditions, rely on different data sources and inventory methods, model different PV technologies at different locations, and consider different lifetimes and analytical periods"* (p. 3316), which ultimately results in a broad range of obtained values. The sensitivity analysis had shown that the variation in results can be considerable even if the data source and boundary conditions are the same.

Literature reviews on the topic have been compiled by Pacca et al. (2007), Sherwani et al. (2010), Sumper et al. (2011) and most recently Peng et al. (2013) covering LCA studies from 1989 to 2011. Within these reviews c-Si PV systems account for EPBTs between 1.7 and 12 years and carbon footprints from below 10 to over 200 gCO_2eq/kWh. In case of CIGS, EPBTs from 1.4 to 2.9 years and carbon footprint from 21 to 95 gCO_2eq/kWh were found in literature (cf. Peng et al. 2013). Most authors studied small-scale rooftop mounted systems or large PV parks. Results for a more diverse range of PV applications were reported by Jungbluth et al. (2005; 2010). In the following, literature data will be compared to the findings of the present LCA first with a focus on module technologies and afterwards with emphasis on different installation options.

In order to limit the range of results and account for more recent technological developments Peng et al. (2013) summarized the results of PV LCA studies from 2005 on and normalized them to an insolation of 1,700 $kWh/m^2/year$ and a lifetime of 30 years as formerly reported by papers of de Wild-Scholten (2011), Raugei et al. (2007), Alsema et al. (2006) or Fthenakis and Alsema (2005). The literature results for the EPBT and carbon footprint of Mono-Si, Multi-Si and CIGS technology can be found in Fig. 3.10. The figure additionally depicts the range of results found in this LCA study. Since the reference irradiation used by Peng et al. (2013) is approximately the one of Cadiz (Spain), the

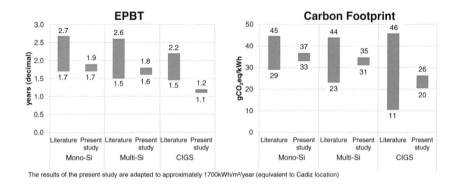

The results of the present study are adapted to approximately 1700kWh/m²/year (equivalent to Cadiz location)

Fig. 3.10: Literature result ranges of EPBT and carbon footprint in comparison to the range found by the present LCA; Source: Authors, adapted from Peng et al. 2013

"present study" values refer to the obtained indicator results for this location.[15] The results from the present study are based on installation type differences of slanted roof, flat roof and open ground systems. Facade systems are not accounted for since the literature review of Peng et al. (2013) addresses them neither. It can be seen that EPBT of Mono-Si and Multi-Si are within the observed literature values but at the lower boundary. They come closest to the EPBT of approximately 1.75 years found de Wild-Scholten (2011) respectively. The obtained EPBT results for CIGS modules are lower than those observed in literature. This is mainly attributable to the underlying renewable energy mix for CIGS module production at Solibro. If an adapted electricity mix were used the EPBT would probably be prolonged to the lower boundary of the literature results as reported by de Wild-Scholten (2011). In terms of carbon footprint the present results are located approximately in the middle of the literature range with the highest similarity to the studies of Alsema and de Wild-Scholten (2006) and de Wild-Scholten (2011) who calculated 30 to 35 gCO_2eq/kWh for Mono- and Multi-Si rooftop systems. The carbon footprint of CIGS modules is slightly lower as the results of

---

[15] A complete EPBT and carbon footprint indicator results table for Cadiz can be found in Tab. A.9 in the Annex.

de Wild-Scholten (2011) where about 30 gCO_2eq/kWh are reported for on-roof installations.

The compilation by Peng et al. (2013) and former literature reviews (e. g. Sumper et al. 2011 or Pacca et al. 2007) are mostly based on either smaller rooftop systems or large scale PV parks, hence, the facade installations studied here are lacking literature backup. Estimates are available from Jungbluth et al. (2010), i. e. from the EcoInvent 2.2 database. The authors report rather high EPBT of facade systems ranging from 4.1 to 4.9 years based on an insolation of 1,117 $kWh/m^2/year$ which is quite similar to the irradiation in Munich. Jungbluth et al. (2010) only account for the non-renewable share of the CED, i. e. apply a differentiated EPBT calculation method which results in systematically lower values. Facade system addressed in this survey show a faster energetic amortization of 3.3. to 3.7 years in the Multi- and Mono-Si case and 2.1 to 2.2 years in the CIGS case.

Large scale open ground systems receive particular attention because of their recent and rapid worldwide expansion (cf. Beylot et al. 2012; Turney and Fthenakis 2011). Among the investigated environmental impacts are primarily energy and climate change related impacts, i. e. the EPBT and carbon footprint, but also land use and conversion, human health, wildlife and geo-hydrological resources (cf. Turney and Fthenakis 2011). Large scale PV systems can be also be differentiated from smaller roof systems because of their increased installation efforts and higher material requirements (e. g. electrical equipment, fences or access roads) (cf. Halsah et al. 2013). Ito et al. (2010) studied very large scale PV systems from 10 to 100 MWp capacity, deploying both c-Si and CIGS modules under an insolation of 1,702 $kWh/m^2/year$[16]. They found EPBTs of 2 years for Mutli-Si, 2.6 years for Mono-Si and approximately 1.75 years for CIGS large scale PV systems and thus show slight deviation from the present results for Multi-Si and CIGS but diverge considerably in the case of Mono-Si. Carbon footprints of Ito et al. (2011) are almost equivalent to those found here when using Multi-Si panels but

---

[16] Comparable to the irradiation of Cadiz (Spain) (Tab. A.8) for which EPBT and carbon footprint results can be found in Tab. A.9 in the Annex.

diverge by about 10 gCO_2eq/kWh for Mono-Si and CIGS modules. In a more recent study Beylot et al. (2012) reported a carbon footprint of 37 gCO_2eq/kWh for a Multi-Si system with comparable boundary conditions, suiting the results derived in the present LCA.

In PV LCA literature only the paper of Halasah et al. (2013) considers, to the current knowledge of the writers, an assessment indicator similar to the $NEA_{space,LT}$ which they call "CO_2 offset". Halsah et al. (2013) compare on-roof installations to open ground systems deploying different module technologies. The authors find that open ground systems count with a smaller $NEA_{space,LT}$ than rooftop installations and that Mono-Si modules, possessing the highest yield/space ratio, can achieve higher $NEA_{space,LT}$ values than Multi-Si or CIGS powered PV systems. These aspects are mirrored in the $NEA_{space,LT}$ results of Section 3.3.3. However, Halsah et al. (2013) consider reference PV systems installed in a desert with an insolation of 2,150 $kWh/m^2/year$ which is in the range of results that were obtained in the sensitivity analysis for a system installed in Las Vegas (compare Tab. A.8). They obtained values from 6.3 to 6.4 tCO_2/m^2 for CIGS modules, 7.1 to 7.3 tCO_2/m^2 for Multi-Si and 7.4 to 7.6 tCO_2/m^2 for Mono-Si modules with the lower boundary corresponding to the rooftop system and the upper one to the open ground system. It can be seen that open ground and rooftop systems show only small deviations. This is contrasted to the present findings (see Fig. 3.9) where slanted roof systems can achieve almost double $NEA_{space,LT}$ than open ground systems. The amplitude of deviation between the module technologies and the amount of CO_2eq emission avoidance is comparable to what Halsah et al. (2013) reported.

In summary, it could be shown that the results obtained here mostly represent the lower threshold of what other scholars reported. This could be attributable to the up to date production data used in this LCA study as well as to the manufacturing conditions of Hanwha Q CELLS and Solibro (e. g. using renewable electricity for production, see Section 3.4.3) and their corresponding supply chain. In general the more recent literature data, as shown in Fig. 3.10, could back up the environmental indicators developed. Therefore it seems justifiable and legitimate to expand the LCA results from a micro or producer level to macro level by comparing them to the EEG PV FIT scheme.

4 Environmental Performance of the EEG Feed-in Tariffs for Photovoltaics

From the analysis of the EEG development, its current state and the reasoning behind its FIT scheme (Chapter 2) it could be shown that there is no intentional orientation of the remuneration tariffs on environmental efficiency criteria. This chapter aims at combining and comparing the FITs received by a hypothetical investor/plant operator for different PV systems with the indicator results obtained in Chapter 3. It therefore sheds light on the research question asking if the EEG PV support scheme is (coincidentally) reflects environmental criteria with the investment incentives it provides. To harmonize the indicator results of carbon footprint, $NEA_{space,LT}$ and EPBT they will at first be normalized and checked for statistical correlation (Section 4.1). Section 4.2 will present the framework used to make the EEG's FIT scheme comparable to the environmental indicators. Both sections will then be connected and interpreted in Section 4.3 and Section 4.4.

4.1 Normalization of Indicator Results

The indicator results are normalized through a data transformation. This transformation step is necessary in order to reveal the relative differences among the studied PV installation types and technologies and to create a dimensionless scale to which the EEG support can be compared more easily. Consequently, the data presented in Tab. 3.3, Tab. 3.4, Tab. 3.5 is used and normalized to a scale from 0 to 100 for each indicator. Thereby 100 represents the best result for a specific indicator and

0 the worst. The transformation is carried out by applying the Formula (4.1):

$$NV_{i,j} = \frac{V_{i,j} - V_{worst}}{V_{best} - V_{worst}} * 100 \qquad (4.1)$$

where i indicates the specific installation option (rows in the result matrices) and j the technologies (columns in the result matrices). $NV_{i,j}$ is the normalized value on a scale from 0 to 100, $V_{i,j}$ the current value from the result tables, V_{worst} the minimum value and V_{best}, accordingly, the best value. Because of the different measuring directions of the indicators, carbon footprint and EPBT improve the lower the numerical value is and the $NEA_{space,LT}$ does it with higher numerical values, the transformation with Formula (4.1) additionally leads to a harmonization of the normalized indicator results, so that, independently from the initial direction of measurement, now 100 always represents the best and 0 the worst result. The normalized results can be found in Tab. A.10 in the Annex.

As mentioned above, for further investigations and comparisons, the differences among installation options and technologies within the obtained results are the subject of analysis and therefore more important than the absolute indicator results. Out of this, the normalized results are checked for statistical correlation. In Section 3.3 it could already be indicated that carbon footprint and EPBT results indeed differ in their absolute dimension, unit of measurement and addressed issue, but show the same behavior in terms of the relative difference between installation/technology combinations. This assumed correlation is proven in the upper part of Fig. 4.1 where the normalized carbon footprint and EPBT are plotted against each other. For reasons of clarity the normalized indicator results of CIGS L and CIGS S are averaged since their difference is only marginal (see Tab. A.10). The linear relationship of all 24 installation/technology combinations considered can be seen clearly in Fig. 4.1. The figure indicates that an installation option with a specific module technology which performs well (close to 100) in terms of the EPBT also counts with a good carbon footprint. In fact, the statistical correlation factor is 0.99, signifying an almost prefect linear relationship. Moreover, it is possible to identify three groups of results in the figure. In the lower left part facade solutions deploying c-Si modules are allocated. CIGS solutions can all be found on the upper right part of Fig. 4.1 and therefore count with the best results. The remaining roof and open ground solution are arranged in the middle ranging from 40 to 65. Because of the almost perfect correlation between carbon footprint and EPBT they have the same meaningfulness for the further assessment in this study which implicates that all conclusions made for the carbon footprint can be transferred to the EPBT and *vice versa*. Consequently, both indicators are analyzed together in the upcoming sections.

In the lower chart of Fig. 4.1 the normalized carbon footprint results are plotted against those of the $NEA_{space,LT}$. Since the $NEA_{space,LT}$ entails an area based reference unit the former correlation of the results is eliminated. The statistical correlation factor amounts to -0.09 which indicates no linear interconnection of results, i. e. a PV plant with a good carbon footprint performance does not necessarily entail a high $NEA_{space,LT}$. Especially noteworthy are the PV plants deploying CIGS modules because all count with good carbon footprint results but relatively low $NEA_{space,LT}$ performance. This observation is attributable to their relatively low electric efficiency and consequently high space occupation per unit of installed capacity. For this reason flat roof (normal) systems using CIGS modules perform best in the carbon footprint, but once the space is limited and the installation design gains importance it performs worst in terms of the $NEA_{space,LT}$.

The revealed inconsistency in the indicator results between carbon footprint/EPBT and $NEA_{space,LT}$ challenges a comparison to the EEG support scheme because interpretations and possible implications on the scheme will differ depending on the underlying environmental indicator. In order to ensure a meaningful discussion **??** will critically reflect the environmental criteria under consideration and develop an approach to cope with the ambiguities.

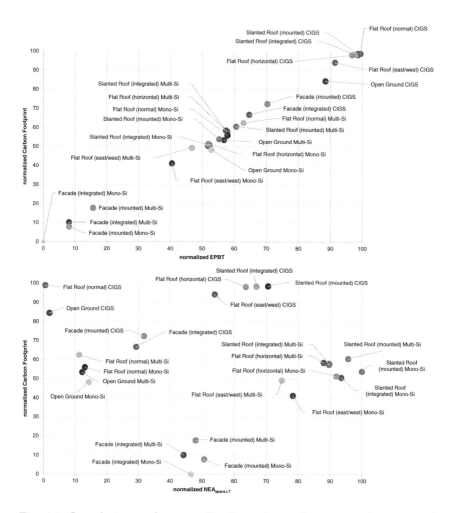

Fig. 4.1: Correlation of normalized carbon footprint, EPBT and $NEA_{space,LT}$ indicator results; Source: Authors

4.2 Determination of a Base for Comparison

This section develops a methodology and base for comparison for the subsequent analysis. It first operationalizes the EEG PV FIT scheme by introducing a set of assumptions (Section 4.2.1) from which afterwards the investment incentives for a set of reference PV systems are derived (Section 4.2.2). The results are from this section enter the comparison to the environmental indicators in the upcoming sections.

4.2.1 Assumptions for the Subsequent Analysis

For a comparison of environmental indicators with the EEG FIT scheme two approaches seem conceivable. First, the comparison could be based on the expected rate of return an investor faces when deciding to install a specific PV plant. Reichmuth (2011, p. 214) states that, given a cost covering remuneration, it is the rate of return which provides the investment incentive in a PV system. Thus, the PV installation offering the highest returns on investment would be preferred by an investor over others. By applying this approach a detailed set of cost data is needed for all 24 technology/installation combinations considered here (compare Fig. 4.1). With the cost data it would be possible calculate the annuity of all life cycle costs and relate it to the specific plant yield in order to obtain the LCOE (cf. Schmidt 2012; Formula (2.1)). The LCOE are comparable to the applicable FIT granted by the EEG in the respective year and for the respective plant under consideration. The difference between FIT and LCOE would consolidate the investor's profit. However, since cost data is highly case specific and dynamic, a comparison based on the observed rate of return for the technology/installation combinations assessed would be required for each year and ideally obtained from a set of case studies instead of a single source. Since this data is difficult to assemble and the determination of the "right level of support" is out of scope in this study, a more theoretical approach is chosen for the subsequent analysis, which will be outlined in the following paragraphs.

As already mentioned in Section 2.3 the aim of a FIT scheme, oriented on the cost of electricity generation, should be to determine an appropriate level of support that neither overcompensates the recipient nor underestimates the required remuneration for an economic amortization. If the tariff is set too high, the social costs of the FIT system are as well and if the tariff is set too low it could not provide an investment incentive in the renewable energy technology (cf. Mendonca et al. 2010, p. 19). Thus, in a perfect world, the observed annual remuneration granted by the FIT scheme (FIT rate ($€/kWh$) multiplied with the reference electricity yield (kWh)) should be (1) at least completely cost covering, i.e represent the annuity of the investment and (2) provide an appropriate rate of return for the investor which is higher than the expected profitability of non-renewable energy projects (cf. Mendonca et al. 2010, p. 19). To optimize the EEG towards this state is one of the main aims of the EEG progress reports (cf. Schmidt 2012; BMU 2007, p. 9). Staiß et al. (2007, pp. 256-279) and Reichmuth (2011, pp. 76-123) for instance carry out a detailed economic analysis along the PV value chain and the end customer sector in order to attain an investment appraisal based on a representative set of PV plants according to Formula (2.1). Keeping this in mind, it is assumed that the observed FIT levels in each year granted by the EEG would provide cost coverage and an equivalent rate of return for PV plant operators, implying an equivalent investment incentive among all PV installation options given the framework conditions of the progress reports (e. g. financing structure, rate of return, time frame or reference yield of the PV plant etc.). Unfortunately, one aspect in the EEG FIT scheme undermines these assumptions: the legislator aimed at setting a PV FIT level that favors building mounted PV installations over those installed on open ground area (compare to Section 2.3). It was previously argued that the extend of this "favoring" is unclear and not observable in the economic analysis carried out by the EEG progress reports (cf. BMU 2007, p. 126; Reichmuth 2011, p. 264). The bonus however, would add some extra remuneration on top of the LCOE of building mounted plants and thus increase their investment incentive over the one for open ground plants. For the analysis carried out here this aspect will be neglected since the extent of the effect is unclear, but it will be kept in mind that open ground plants might be overvalued in the current analysis.

Formula (4.2) recalls the LCOE determination methodology used by the EEG research reports in order to identify the level of support (cf. Schmidt 2012). It is a simplified version of Formula (2.1).

$$FIT \hat{=} LCOE_{ref} = \frac{\text{Annuity of Life Cycle Generation Cost (€)}}{ASAY_{ref}(kWh)} \quad (4.2)$$

The formula expresses the assumption made above: the FIT would equal the $LCOE_{Ref}$ determined with the reference conditions of the EEG research reports and include an equivalent rate of return (considered in the calculative interest rate) (cf. Schmidt 2012, p. 6-8; Staiß 2011, pp. 110-112). The nominator is the annuity of all observed costs and would equal the $LCOE_{ref}$ multiplied with the average annual reference yield ($ASAY_{ref}$) assumed in the EEG research reports. Reichmuth (2011, p. 111) for instance states $900kWh/kWp$ for building installations and $950kWh/kWp$ for open ground PV plants in the year 2011. When introducing the set of installation options studied in this survey a differentiated, actually generated, amount of electricity can be found at the reference location of Munich ($ASAY_{act}$). Accounting for these differences would lead to a varying annual remuneration received by the plant owner which is depending on the respective PV technology/installation option (see Tab. A.3 in the Annex). The following two formulas represent this fact:

$$FIT * ASAY_{ref} = Rem_{ref} \quad (4.3)$$

$$FIT * ASAY_{act} = Rem_{act} \quad (4.4)$$

where Rem_{ref} and Rem_{act} represent the reference annual remuneration received under the electricity yield conditions as defined by the EEG research reports and the actually obtained amount of electricity from a PV plant in this study respectively.

With the help of Formula (4.2), Formula (4.3) can also be rewritten to emphasize the cost and rate of return covering character:

$$LCOE_{ref} * ASAY_{ref} = \text{Annuity of Life Cycle Generation Cost} \quad (4.5)$$

where it can be seen that Rem_{ref} covers the annuity of all life cycle costs and additionally the annual return that can be expected under reference conditions. Given this assumption, the observed differences between Rem_{act} and Rem_{ref} (ΔRem or annual net remuneration) would consolidate the incentive to choose one installation option over another one, or in other words, would increase the returns from installing a certain PV system. This is displayed in Formula (4.6) and Formula (4.7):

$$\Delta Rem = Rem_{act} - Rem_{ref} = FIT * (ASAY_{act} - ASAY_{ref}) \quad (4.6)$$

$$\Delta Rem = Rem_{act} - \text{Annuity of Cost} \quad (4.7)$$

However, it has to be acknowledged that the methodology of determining the level of support is also based on a set of constraining assumptions which could be questioned as such (see Section 2.3). A more differentiated LCOE analysis in the EEG progress and accompanying research reports could probably reveal a diversified portfolio of LCOEs for different installation options and PV technologies deployed and thus undermine the assumption that the observed FIT rate provides equivalent investment incentives under $ASAY_{ref}$ conditions. Moreover, if the bonus granted to building PV systems were known it would lead to a higher ΔRem of those systems in comparison to open ground.

When deriving the investment incentive for a specific installation option in this way it is necessary to be aware of two conceivable interpretation possibilities. The investment incentive could be addressed from a micro or a macro perspective. The former refers to the view of an investor who might be endowed with a certain type and amount of capital and building/open ground area on which the PV plant could be installed. Reichmuth (2011, p. 19) shows that the two largest groups of investors in EEG remunerated PV installations in 2010 were private (38.3%) and commercial/industrial (32.7%) followed by agricultural (25.4%) and public investors (3.5%). When exemplary imagining a private house-owner who decides to equip her residential building with a PV system, it can be presumed that her ability to choose from different installation options is constrained. It depends for instance on the available roof type, i. e. flat roof or slanted roof or if the facade is suitable for a PV system or not. This would also be applicable for owners of production halls with a high potential for large flat roof PV systems or agricultural cooperatives with a certain roof space or open ground space available. However, these examples show that, from a micro perspective, the investor would choose the PV system that maximizes her return on investment, which is in turn constrained by certain prerequisites. Thereby the investor would not necessarily consider that an installation option which is unsuitable for the planned project would, in theory, offer higher returns per installed unit of system capacity. The higher rate of return could for instance be induced by deviations in ΔRem[1] but also from an environmental premium (see Section 5.2). This might be different in case of investment firms that are not facing such constraints and seek for the optimal conditions to maximize the return on investment. Nevertheless, the outlined argument captures the diversity of possible constraints and prerequisites on the micro level.

On a macro scale, it could be argued that these diverse constrains level out and the PV capacity expansion will be oriented on the installation options that offer the highest investment incentive. This argument is also pointed out by Reichmuth (2011, p. 214) who states that, if a cost covering support is provided, the amount of capacity enlargement essentially depends on the rate of return received by the investor. Following the assumptions made above, this differing rate of return would be consolidated by ΔRem in $€/kWp$.

---

[1] In practice Rem_{act} can also deviate because of observed cost differences since, from a micro perspective, different framework conditions are prevailing (e. g. financing conditions).

4.2.2 Determination of Reference FITs

In order to operationalize the assumptions made in the preceding section the FITs granted by the EEG are derived for all specific installation/technology combinations displayed in Fig. 4.1. As the analysis of Section 2.2.2 has shown, the average FITs received by a plant owner for feeding in PV electricity to the grid are, since the 2004 amendment, dependent on the operating capacity of the specific PV plant (among other factors). Unfortunately, because installation sizes are case dependent, the received average FIT rates cannot be generalized and constraining assumptions have to be introduced. Therefore, the eight installation types studied in the LCA are considered in three respective size classes that capture the stepped tariff design implemented in the EEG. It should be noted that the stepped tariff design does not necessarily contradict the assumptions made above because the size classes are introduced with the intention to reflect size dependent cost differences. Thus, when comparing PV plants with equivalent sizes the annuity of cost should also be equivalent. In terms of environmental performance it was assumed that the indicator results are linearly scalable (see Section 3.2), i. e. the PV plant size is considered to be roughly proportional to the environmental impact. However, because in practice some sizes of installation options are unrealistic, e. g. a $5kWp$ open ground plant or a $20MWp$ facade solution, a set of reference plants is developed for which average FITs are determined.

The EEG (2000) remunerated all types and sizes with the same FIT except open ground plants greater $100kWp$ that were excluded from the scheme. From 2004 on, the size dependent stepped tariff was introduced for building PV systems and comprised a differentiation in installed capacity shares smaller $30kWp$, between 30 and $100kWp$ and greater than $100kWp$ linked with respectively decreasing FIT payments (cf. § 11 EEG 2004). In the 2009 amendment another remuneration segment for capacity shares greater $1MWp$ was introduced (cf. § 33 EEG 2009). Finally, in the revision that entered into force in April 2012, the stepped tariff was adapted again and currently comprises a differentiation in ca-

Tab. 4.1: Reference PV plant sizes considered for FIT determination; Source: Authors

Reference plant size in kWp	Small	Medium	Large
Slanted Roof (mounted)	10	100	200
Slanted Roof (integrated)	10	100	200
Flat Roof (normal)	50	500	1,500
Flat Roof (east/west)	50	500	1,500
Flat Roof (horizontal)	50	500	1,500
Facade (mounted)	10	100	200
Facade (integrated)	10	100	200
Open Ground	1,000	10,000	20,000

pacity shares until $10kWp$, from 10 to $40kWp$, from $40kWp$ to $1MWp$, from $1MWp$ to $10MWp$ and greater $10MWp$ (§ 32 EEG, 2012-PV). Under the EEG (2012-PV) open ground plants received a uniform tariff up to $10MWp$ and no remuneration beyond this capacity.

In order to represent all these size classifications, the reference plant capacities for the corresponding installation options are chosen as outlined in Tab. 4.1. Moreover, the sizes were oriented on experiences of Hanwha Q CELLS (cf. e. g. Hanwha Q CELLS n.d.) and partially based on the elaborations of Staiß et. al. (2007, p. 264) and Reichmuth (2011, p. 106). For each of the outlined cases the FIT is determined according the EEG framework conditions (EEG 2000; 2004; 2009; 2012 and 2012-PV) at the beginning (1st January) of the year 2001, 2004, 2009, 2012 and 2013 as described in Section 2.2.2 (see Tab. 4.2). Subsequently, the resulting nominal FITs[2] for small, medium and large reference plants of are arithmetically averaged in the specific year in order to obtain a single FIT for each installation type that represents the stepped tariff design of the EEG. To be consistent with the assumptions made above the average FIT value shall represent the average annuity of cost of all size classes considered divided by the reference electricity yield ($ASAY_{ref}$). If exact data on size classes and installed system types were available, the arithmetical mean used here could also be replaced by a FIT composed of a weighted sum of the different size classes actually installed in a respec-

---

[2] Nominal FITs are measured in terms of prices of the respective year under consideration.

Tab. 4.2: EEG FIT rates of different installation types; Source: Authors based on EEG (2000, 2004, 2009, 2012, 2012-PV)

	Installation type	Size (kWp)	FIT (ct/kWh)				
			2001	2004	2009	2012	2013
Building	Slanted Roof (mounted),	10	50.62	57.40	43.01	24.42	17.02
	Slanted Roof (integrated),	100	50.62	55.44	41.54	23.59	13.84
	Facade (mounted),	200	50.62	54.72	40.56	22.78	13.40
	Facadace (integrated)	Average	50.62	55.85	41.70	23.60	14.75
	Flat Roof (normal),	50	50.62	56.28	42.17	23.94	14.71
	Flat roof (east/west),	500	50.62	54.29	39.97	22.30	13.14
	Flat roof (horizontal)	1,500	50.62	53.92	39.58	22.01	12.63
		Average	50.62	54.83	40.57	22.75	13.49
Open Ground	Open Ground	1,000	5.06	45.70	31.94	18.35	11.78
		10,000	0.51	45.70	31.94	18.35	11.78
		20,000	0.25	45.70	31.94	18.35	5.89
		Average	1.94	45.70	31.94	18.35	9.82

tive year. Besides this aspect, the facade bonus of § 11 (1) EEG (2004) and the market integration model introduced by § 33 (1) EEG (2012-PV) are included in the assessment. However, for reasons of simplicity and to avoid further constraining assumptions the own use option that was applicable from 2009 until 2012 as well as the market premium model for direct PV electricity marketing are not addressed in the following analysis. In case of the open ground PV systems it is assumed that all plants considered are installed on eligible areas and therefore receive an EEG remuneration. Differentiated FITs according to the land type used for open ground installations (e. g. § 32 (3) EEG 2011) were also averaged since the impacts of a specific land use type are not addressed in this study. The facade bonus is not included in the FIT values of the table because it cannot be considered as a component of the FIT in its strict sense (representation of $LCOE_{ref}$) but will be addressed below.

As it can be seen in Tab. 4.2, open ground plants receive the lowest average FIT per generated kWh electricity. Generally, the simple EEG categorization of the FITs in building and open ground installations accounts neither for installation nor technological differences within these categories. The slanted roof, facade and flat roof variants considered in the present analysis would therefore all be remunerated under the uniform conditions of the building category. FITs of slanted roof and facade installations in comparison to flat roof solutions differ slightly because of the stepped tariff in combination with the underlying plant size assumptions. Independently from the module technology chosen, FITs remain constant.

In a next step Formula (4.6) is applied by using the FITs of Tab. 4.2, the $ASAY_{act}$ of Tab. A.3 in the Annex and the reference yield ($ASAY_{ref}$) defined by Reichmuth (2011, p. 111) for the year 2011.[3] The resulting figures for ΔRem in €/kWp can be found in Tab. A.11 in the Annex. Some of the values for ΔRem are negative since their $ASAY_{act}$ is lower than in the reference case. A negative ΔRem could therefore either lower the rate of return scheduled by Reichmuth (2011) or even lead to a remuneration that is not cost covering. However, since the exact costs are unknown it can still be concluded that, the lower ΔRem, the lower the investment incentive, either because costs are not recovered or the rate of return is insufficient. Moreover Tab. A.11 reveals higher investment incentives in CIGS than in c-Si systems since they provide higher electricity yield per installed unit of capacity.

The analysis carried out only provides a snapshot approach based on 2011 technology, i. e. the environmental data, the $ASAY_{act}$ as well as the $ASAY_{ref}$ values of Reichmuth (2011, p. 111) refer to the year 2011. To avoid the pitfall of comparing differing absolute FITs in the respective years, the values for ΔRem are, in a last step, normalized according to Formula (4.1), where V_{best} refers to the highest and V_{worst} to the lowest ΔRem and consequently the highest and the lowest investment incentive. Unfortunately, the emerging data is still very complex (see Tab. A.12 in the Annex) and unsuitable for a proper comparison. The normalization has the advantage of focusing on relative instead of absolute differences which prove to be significantly varying between installation options but are rather stable among the considered module

---

[3] It should be noticed that the described data transformation relates FIT rates that were determined based on past technologies and investment costs to electricity yields of current PV technologies. This aspect would be highly problematic when comparing the absolute remuneration values as displayed in Tab. A.11 among each other because the electrical efficiency of PV modules and BoS components increased and the respective investment cost decreased from the year 2000 until nowadays. Therefore it was decided to compare the normalized data focusing on relative differences.

technologies. Because of this the normalized values of different technologies are averaged to reduce complexity (see Tab. A.12 in the Annex). This set of data transformations operationalizes the FIT schemes laid out by the EEG amendments and builds the basis for a comparison to the determined normalized carbon footprint/EPBT and $NEA_{space,LT}$ results among the considered installation options. Fig. 4.2 summarizes the normalized ΔRem values used for the following assessment.

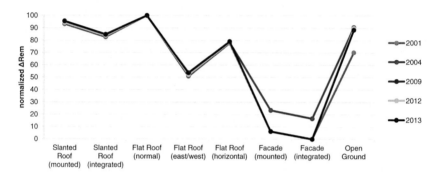

Fig. 4.2: Normalized ΔRem for PV installation types under different EEG FIT schemes dismissing technological differences; Source: Authors

Normalized values have to be interpreted necessarily in relation to each other, i. e. the worst and best value within a specific EEG FIT scheme determine the upper and lower boundaries of 100 or 0. Consequently, a change in these boundary values implies a change of all other normalized values obtained. Additionally, it should be considered that a normalized value of 0 does not necessarily indicate no ΔRem. The integrated facade solution for instance counts with a value of 0 under all EEG FIT schemes except 2004, despite the fact that its average FIT (see Tab. 4.2) is one of the highest granted in the respective year. However, because of its actually low electricity yield ($ASAY_{act}$), significantly deviating from the reference yield ($ASAY_{ref}$) defined by Reichmuth (2011, p. 111), it counts with the lowest investment incentive and is represented by the normalized value of 0. The flat roof (normal) solution in turn exceeds the reference yield the most leading to the highest return and thus highest normalized value (100).

In general the graphs resemble each other quite well. Flat roof (normal) systems would provide the highest investment incentives, interpreted on a macro scale and given the above-mentioned assumptions, followed by open ground plants, the slanted roof solutions and the two other flat roof alternatives. Facade systems count with the comparably lowest returns. Deviations can be found for the open ground system in 2001 and the facade systems under the 2004 framework conditions. The EEG (2000) granted the same FIT for all PV installation types besides open ground systems which were only remunerated up to $100kWp$. Since the open ground reference plants addressed are significantly larger, the FIT value used to calculate the ΔRem was, in relative terms, lower than in the subsequent years. It is indeed questionable if the very low average FIT for open ground plants granted by the EEG (2000) (see Tab. 4.2) is cost and rate of return covering. Therefore, this deviation will not be further evaluated in the following paragraphs. Moreover the already mentioned problem of the undetermined bonus for building PV systems biases the normalized values. It underestimates the investment incentive in all building installation options and thus favors the ΔRem value for the open ground system. When interpreting the lines of Fig. 4.2 it is important to notice that first of all the order or succession determines if one PV installation option would be predominantly installed. Secondly, also the magnitude of relative differences is important to consider for assessing efficiency because it could be re-translated in an absolute amount installed capacity.

The increased performance of facade systems under the 2004 EEG scheme is particularly interesting since the deviations are attributable to the, at the time granted, facade bonus of $5ct/kWh$. This bonus or premium was introduced to compensate the higher LCOE and to provide additional incentives to deploy the high spatial potential of facade options (cf. EEG-Explanation 2004, p. 41). The facade bonus is granted on top of the FIT and therefore enters the calculation of Formula (4.6) separately. Section 5.2 discusses the effects and calculation rationale of such a bonus option.

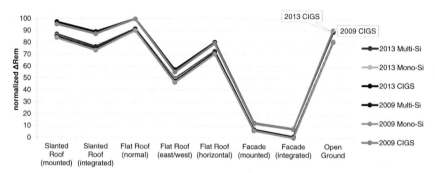

Fig. 4.3: Normalized ΔRem of PV installation options accounting for technological differences; Source: Authors

Accounting for differences in technologies requires a normalization among all ΔRem values within a respective year (see Tab. A.13 in the Annex) and leads to an adapted base for comparison, depicted in Fig. 4.3. The normalized values are now deviating depending on the PV module technology chosen and are consequently not averaged. Because of their higher electricity yield per unit of installed capacity CIGS modules outperform c-Si modules independently from the installation option chosen and would entail an increased investment incentive, *ceteris paribus*. The different amendments of the EEG do not significantly influence the trend of the lines in Fig. 4.3, that is why the values of 2009 and 2013 are exemplary shown. In further comparisons only the 2013 line for CIGS, Mono- and Mutli-Si will be displayed as a representation of the support scheme.

4.3 Climate Protection and Primary Resource Consumption

In a next step, the normalized annual remunerations differences from Fig. 4.2 are combined with the normalized indicator results for the carbon footprint and EPBT presented in the upper chart of Fig. 4.1. Since the environmental analysis, based on the LCA, focused on comparing the potential environmental impacts from both, different installation options and different PV module technologies, these aspects are separately assessed as well. At first, the EEG support is compared to the environmental performance of the eight considered installation options dismissing the technological differences. This is done by normalizing the carbon footprint/EPBT indicator results for Mono-Si, Multi-Si and CIGS separately among the installation options; i. e. the normalized value of 0 refers to the worst indicator result of a certain technology and 100 represents the best result for the same technology.

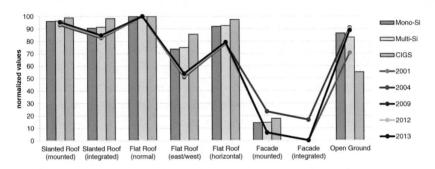

Fig. 4.4: Technology specific normalized carbon footprint/EPBT (bars) in comparison to normalized ΔRem under different EEG FIT schemes (lines); Source: Authors

As it can be seen in the blue, gray and green bars of Fig. 4.4 the normalization almost eliminates technological variations. The only exemption is the open ground CIGS system whose relative performance is worse than the one of the Mono-Si and Multi-Si open ground system. This can be explained by the diverging system design (see Section 3.2 and Tab. A.1 in the Annex) but will be neglected in the further deliberations. At a first glance, it can be observed that in general installation types with a good environmental performance (close to 100) entail a higher investment incentive than those with a low environmental performance. The match is to a large extent attributable to the dependence of both data on the annual electricity yield ($ASAY_{act}$) outweighing the diverging environmental impacts related to one unit of installed capacity. Those installation options with a high yield count with an improved carbon footprint (Formula (3.2)) and EPBT (Formula (3.1)) and at the same time receive comparably higher remunerations than in the cost and rate of return covering reference case.

A particularly good match of flat roof (normal), slanted roof (mounted and integrated) with the ΔRem lines is identifiable, indicating that the installation options with the best environmental performance would be the most profitable to invest in. In case of flat roof (east/west and horizontal) the relative differences in ΔRem are underestimating the environmental performance in terms of carbon footprint and EPBT. For facade solutions, the lines representing the EEG framework conditions also reflect the inferior environmental results. Since facade solutions count with a low $ASAY_{act}$ their ΔRem as well as their carbon footprint/EPBT are menial. As mentioned above, the facade bonus granted by § 11 (2) EEG (2004) increases the investment incentive in facade systems. The extent of the bonus, however, would overcompensate the integrated facade solution in relative terms. Open ground systems receive a relatively high support in comparison to their environmental impact. This is due to their high yields and the biased ΔRem values (see above).

It can be concluded that generally PV installation options with higher electricity yields resulting from either higher electrical efficiencies, a better tilt towards the sun or higher location dependent insolation are more likely to be chosen by investors because of higher returns and at the same time count with improved EPBTs and carbon footprints. Mismatches in support could be identified for flat roof (east/west) and flat roof (horizontal) installations which are currently slightly undervalued by the EEG. Thus, the results could point at an insufficient differentiation in the EEG remuneration for roof mounted systems.

The picture changes when acknowledging technological differences between Multi-Si, Mono-Si and CIGS modules. As Fig. 4.5 shows PV systems equipped with CIGS modules outperform c-Si systems in all installation types. Multi-Si systems show slightly better results than Mono-Si systems because of their less energy and carbon intensive production process. The higher investment incentive for CIGS modules based on their higher yield per unit of installed capacity reflects the outcome of carbon footprint and EPBT. Consequently, from the perspective of carbon footprint and EPBT the EEG coincidentally provides investment incentives in the environmentally superior installation or technology, the extent, however, is not mirrored in all cases.

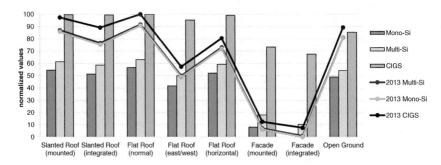

Fig. 4.5: Normalized carbon footprint/EPBT results in comparison to normalized ΔRem accounting for technological differences; Source: Authors

4.4 Space Requirements

Next, the indicator results of $NEA_{space,LT}$ are analyzed in comparison to the observed EEG support in 2001, 2004, 2009, 2012, 2013. Fig. 4.6 depicts the normalized indicator results, not accounting for technological differences, compared with the normalized net remunerations from Fig. 4.2. In the analysis of the carbon footprint and EPBT above it already became clear that the electricity yield is an important factor responsible for matches between environmental performance and support received for a specific PV installation. This is also applicable for the $NEA_{space,LT}$ since the avoided CO_2eq emissions are a function of the generated electricity.

However, the effects induced by the PV system yield are constrained by the space intensity of an installation option. For this reason, the normal flat roof and open ground systems entail by far the worst normalized environmental indicator outcome in Fig. 4.6.

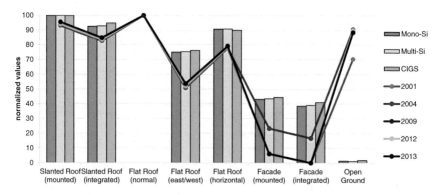

Fig. 4.6: Technology specific normalized $NEA_{space,LT}$ (bars) in comparison to normalized ΔRem under different EEG FIT schemes (lines); Source: Authors

The comparison of the environmental performance to the EEG support shows a good match of the mounted and integrated slanted roof plants with the level of ΔRem in all EEG versions. Both systems require the same amount of roof space, but the integrated system yields a slightly lower amount of electricity which in turn leads to less avoided emissions and also lower annual remuneration. A partial match between the remuneration from 2001 to 2013 and the environmental performance is also identifiable for the flat roof (east/west) and flat roof (horizontal) installations. These systems achieve the third and fourth best $NEA_{space,LT}$ and can count with the fifth and sixth highest net remuneration, respectively. However, from this point of view, these installation options are still undervalued by the respective EEG support. Among the remaining installation options no significant conformity to the support distribution under all EEG framework conditions appears.

Introducing space requirements as a limiting factor in the indicator resulted in flat roof (normal) and open ground systems attaining the worst outcome among the installation options because of their high space losses due to occurring shadings (Section 3.3.3). An improved space utilization could increase the environmental efficiency and also the amount of electricity generated with the same amount of area deployed (compare to the arguments in Section 3.1). The comparison of the $NEA_{space,LT}$ with the EEG framework conditions indicates an overcompensation of the flat roof (normal) and open ground system relative to the other installation options, keeping the overestimation of ΔRem values of open ground system in mind. Especially the former receives the highest relative support from 2001 until 2013 but counts with the relatively worst environmental performance. Facade solutions in turn are underrepresented by the support conditions prevalent in 2001, 2009, 2012 and 2013 in terms of the $NEA_{space,LT}$. Here the line of 2004 shows how an installation specific premium, granted on top of the FIT, can improve the investment incentive in such installations and hence their conformity to the $NEA_{space,LT}$.

In summary it can be found that, concerning installation types, no significant match between remuneration by the EEG and $NEA_{space,LT}$ performance is prevalent besides for slanted roof PV systems. I. e. the support scheme in the versions of 2001 to 2013 does not sufficiently account for space efficiency and climate protection. The omitted differentiation between types of building installations is especially noteworthy.

Accounting for technological differences leads to a diversification of the normalized indicator results and support granted (see Fig. 4.7). CIGS modules perform worse than c-Si modules because of their lower electrical efficiency and higher space intensity. The differences between Multi and Mono-Si modules are marginal. Flat roof (normal) and open ground systems remain significantly overvalued when comparing the investment incentive to the $NEA_{space,LT}$. It could be argued that, independently from the installation option chosen, deploying CIGS instead of c-Si modules in a PV system deteriorates the environmental efficiency from the perspective of the $NEA_{space,LT}$, providing incentives to install rather space intensive PV systems and using high yield locations inefficiently.

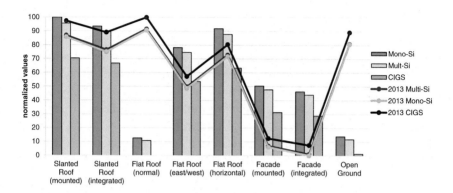

Fig. 4.7: Normalized $NEA_{space,LT}$ (bars) in comparison to normalized ΔRem under different EEG FIT schemes (lines); Source: Authors

When comparing normalized values among each other a clear focus is set on the relative differences between technologies and installation options to reach a higher degree of environmental efficiency in the support scheme. Thereby it is easy to disregard the importance of the absolute level of the environmental indicators. Relative differences among technologies and installations can be equivalent but established on either a high or low absolute level. An environmentally efficient support scheme should therefore also aim to foster a low level of environmental impact in absolute terms. Leaving aside technological and installation specific differences, the sensitivity analysis of the preceding chapter had shown how indicator values can vary with adjusted input variables. If for instance the electric efficiency of a module or BoS parts improves, the electricity yield as well as the space intensity would be enhanced, leading to a better performance in all indicators and implying a lower level of environmental impacts in absolute terms. The same argument would be valid for e. g. improved production processes, environmentally damaging material bans, material substitution, shorter transport routes, etc.

5 Discussion of Results

Based on the findings of Chapter 4 this chapter presents an approach for deriving practical implications from the obtained results (Section 5.1) and subsequently captures the third sub-research question of this study by discussing possible effects on the policy design and related conceivable economic effects when accounting for environmental efficiency criteria (Section 5.2).

5.1 Result Preparation for Decision Making

The preceding chapter showed how certain environmental indicators can be operationalized for a comparison to the EEG framework conditions. The indicators were chosen as an exemplary representation of three criteria that in turn were oriented on the main objectives of the EEG. The analysis was carried out in order to address the issue of environmental efficiency as defined in Section 2.5 and accordingly answer the research question if the FIT system applied by the EEG provides investment incentives that are consistent with the environmental criteria. From the comparison it could be revealed that the research question cannot simply be answered with yes or no. The answer is rather dependent on a set of assumptions and methodological constrains which have to be introduced. Despite the methodological limitations of the carried out comparisons (see Section 4.2) the main issue for approaching the research question is the choice of suitable environmental criteria and consequently indicators and reference units to represent them (compare to the deliberations in Section 3.1).

While carbon footprint and EPBT results suited the support conditions quite well, $NEA_{space,LT}$ results revealed significant mismatches, leading to differing implications for decision making. E.g. the flat roof (normal) installation performed best in terms of EPBT and carbon footprint and simultaneously one of the highest net remunerations. In terms of $NEA_{space,LT}$ instead, this installation option performed worst. Focusing on the carbon footprint and EPBT would lead to the conclusion that the support scheme reflects favorable environmental performances of installation options and technologies and that there would be no urgent necessity for changes in the EEG support towards environmental efficiency. Taking the $NEA_{space,LT}$ results as a decisive criterion implies a relative overcompensation and a possible necessity to revise the remuneration level when intending to base the support scheme on environmental criteria. Hence, an approach to cope with the obtained ambiguous results is necessary in order to reveal meaningful implications for decision making.

"Decision-making is the study of identifying and choosing alternatives to find the best solution based on different factors [...]" (Mateo 2012, p. 7) Multicriteria decision analysis (MCDA) has been frequently applied to cope with the problem of making a meaningful decision out of a possibly conflicting set of criteria (cf. Mateo 2012, p. 7; Madlener and Stagl 2005) and could therefore be a useful tool rank and prioritize the results of carbon footprint, EPBT and $NEA_{space,LT}$ (cf. from Fig. 4.4 to Fig. 4.7). Transferring the concept to the results of this study and exemplary applying two simple MCDA methods, weighted sum and analytical hierarchy process (AHP), enables to rank the installation options according to a specific score. Technological differences are left aside here because of the clear superiority of CIGS over c-Si modules in the case of carbon footprint/EPBT and a reversal of results when considering $NEA_{space,LT}$. Accounting for them would make the MCDA analysis more complex and difficult to interpret. Both MCDA methods need to be based on assumptions about the weight of each goal/indicator. The weights applied in the following are determined according to the methodology lined out by Mateo (2012, p. 12). A pairwise comparison matrix is created and

numerical values from 1 to 9 represent the prioritization of one goal over another, with 9 indicating that an objective in a respective line is absolutely more important than another objective in the respective column and 1 implying an equal importance of objectives (see left side of Tab. 5.1). According to the EEG-Explanation (2009, p. 19) the objective of mitigating climate change is assumed to be slightly more important than environmental protection (value 3) and very strongly more important than resource consumption (value 7). Moreover the goal of environmental protection could be strongly more important (value 5) than resource consumption. This shows that a clear prioritization of goals is necessary in order to apply MCDA comprehensively. Summing up the respective values in each column and dividing the actual value in each cell by the respective column sum results in the values presented on the right side of Tab. 5.1. To determine the weight of each indicator the ratios in each row are averaged (cf. Mateo 2012, p. 13).

Tab. 5.1: Prioritization of indicators and derivation of weights for MC-DA; Source: Authors

Prioritization	Climate change	Environmental protection	Resource consumption		Weights	Climate change	Environmental protection	Resource consumption	Average
Climate change	1	3	7		Climate change	68%	71%	54%	64%
Environmental protection	1/3	1	5		Environmental protection	23%	24%	38%	28%
Resource consumption	1/7	1/5	1		Resource consumption	10%	5%	8%	7%
Sum	1.48	4.20	13.00		Sum	100%	100%	100%	100%

The obtained weights are applied to the normalized indicator results (Fig. 4.4 and Fig. 4.6) by using a relative pairwise comparison framework according to the AHP methodology (cf. Mateo 2012, pp. 13-14). Thereby scores as outlined in Tab. A.14 in the Annex are revealed and can, after being again normalized according to Formula (4.1), be compared to the support scheme as depicted in Fig. 5.1. Slanted roof and flat roof (horizontal) installations perform best and because of the relatively high importance of the $NEA_{space,LT}$ the overall score of flat roof (normal) and open ground installations is considerably low. However, the methodology entails a significant drawback when applying it with normalized values from 0 to 100. In the development of the pairwise comparison matrix, which states the degree of superiority of one installation option in comparison to another, a division by zero is necessary which is mathematically undefined. The relative weight of each installation with a normalized value of 0 within a respective indicator was therefore simply set to zero, indicating that it has no impact on the evaluation end-score.

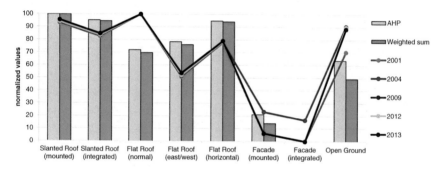

Fig. 5.1: Normalized MCDA scores (bars) in comparison to normalized ΔRem (lines); Source: Authors

Using the determined indicator weights from Tab. 5.1 in a weighted sum method (cf. Mateo, 2012, p. 19) reveals almost the same results than the pairwise comparison AHP method (see Tab. A.15 in the Annex). After a normalization of the weighted sum scores they are comparable to the normalized AHP scores and the net remunerations resulting from EEG support scheme excluding technological differences (Fig. 5.1). The figure reveals a quite good match between aggregated environmental performance and remuneration in the respective years. Flat roof (normal) and open ground installations remain relatively overcompensated, acknowledging the uncertainty in open ground ΔRem values, and flat roof (east/west) and flat roof (horizontal) are undervalued by the remuneration scheme. The facade bonus of the EEG (2004) reflects the environmental performance of the mounted facade solution but would overcompensate the integrated facade solution.

As it could be seen, applying MCDA to the obtained ambiguous indicator results can be fruitful for reducing the complexity and building a basis for subsequent decision making. However, additional assumptions and an explicit political goal prioritization are necessary to aggregate the outcomes. The process of goal prioritization is normative and should therefore involve a variety of stakeholders at the science, policy and public interface (cf. McCool and Stankey 2004). Since the indicators and installation options proposed and exemplary assessed in this survey are limited in number and reveal only one contradiction between the carbon footprint/EPBT and the $NEA_{space,LT}$, their implications for the support scheme will be further discussed separately. This approach can avoid the normative decision on which goal and hence which indicator would be more important to address. At the same time it offers a broader insight on possible policy implications when trying to orient the PV remuneration scheme on environmental criteria. Nevertheless, it could be recommended to apply a MCDA in case that more criteria and indicators are used for an environmental efficiency assessment of the EEG. This would for instance be the case when analyzing the environmental indicator set of UNEP (2010, p. 37) as proposed in the preceding section. Moreover, MCDA could be a beneficial tool when carrying out a more generous goal assessment of the EEG, not only accounting for the environmental objectives but also for costs of electricity supply, technology development and the achievement of capacity expansion (see Tab. 2.2).

5.2 Effects on the Policy Design

Based on the comparison of the environmental indicator results to the EEG support scheme (Chapter 4) this section aims for an analysis of possible consequences on the PV support scheme by addressing the third sub-research question: Which potential improvements can be derived from the analysis and what are their implications for the EEG and PV support schemes in general?

The preceding assessment could derive shortcomings of the EEG support scheme on two dimensions. It was found that the differences in environmental performance of (1) installation options and (2) technological differences are not accounted for and that results and possible implications differ depending on the chosen base for comparison (or environmental indicator). The upcoming discussion will therefore account for those differences by examining conceivable consequences for the EEG FIT scheme. Moreover, possibilities to include environmental aspects in other regulatory PV support options, described in Section 2.1.2 and summarized in Tab. 2.1, are assessed, broadening the scope away from the pure FIT assessment. Limitations in the total technical potential for several PV installation options are not addressed since it is presumed to be sufficient. The UBA (2010, p. 48) for instance states an availability of roof space of $800km^2$, facade space of $150km^2$ and $670km^2$ open ground space in Germany. This availability should not be confused with the arguments made above in favor of indicators for space efficiency like the $NEA_{space,LT}$. Increasing space efficiency will nevertheless contribute to an overall increased environmental performance of the support scheme since highly suitable locations can yield more electricity.

5.2.1 FIT Scheme

In case of FIT schemes, Madlener and Stagl (2005) argue that *"it is basically in the hands of the policymakers (and their skills and motivations) to allow for at least some degree of economic efficiency [...], to steer the diversity of technologies employed (i. e., by widening or narrowing the range of eligible RET), and to have an influence on the environmental and social impact of the RET mix (i. e., by deliberately under- or oversubsidising certain technologies)"*.[1] Consequently, the political sphere is the one able to adapt a FIT scheme and thus influence the investment incentives on a macro scale in accordance to the identified environmental shortcomings.

---

[1] RET abbreviates "renewable energy technology".

When aiming to adapt the fixed price FIT scheme applied by the EEG some of the design options discovered in Section 2.1.4 seem conceivable to account for the environmental performance of different PV installation options and technologies. Generally, those installations or technologies with too low investment incentives from an environmental point of view should receive an increased yearly net remuneration from the FIT scheme relative to others and, on the contrary, those that are overrepresented should receive an effectively lower support. In the following, possible changes to the prevailing FIT scheme are considered. A bonus option is discussed in detail before other FIT design parameters are addressed. Afterwards a possible market premium FIT is scrutinized briefly. At the end of this subsection a short summary of recommendations for conceivable changes in the FIT scheme is given.

First, an additional environmental bonus on top of the cost and rate of return covering fixed FIT could be granted, representing environmental superiority, similar to the facade bonus granted by the EEG (2004) to account for higher LCOE of facade systems (cf. EEG-Explanation 2004). Klein et al. (2008, p. 77) conclude that the extra premiums can be reasonable measures to contribute reaching policy goals. In Section 2.3 it was found that the EEG applies such a bonus since its introduction for PV building systems in comparison to open ground systems. However, unfortunately its extend could not be revealed. Moreover, the law made use of such an option in the support conditions for electricity generation from biomass and geothermal energy in case of using innovative technologies (cf. e. g. §§ 27 (4), 28 (2) EEG 2009; Klein et al. 2008, p. 56). This shows the general acceptance and implementation of such an option within the EEG remuneration scheme. Thus, when trying to implement an environmental bonus for PV installations or technologies to ultimately increase the environmental efficiency it remains questionable how the bonus level should be fixed, which installations or technologies should receive the bonus and what are the consequences concerning economic and environmental efficiency.

From a technical point of view, Formula (4.7) can be adapted to account for the bonus:

$$\Delta Rem_{env} = (FIT + B_{env}) * ASAY_{act} - \text{Annuity of Cost} \qquad (5.1)$$

where the FIT again represents the reference cost and rate of return covering support (i. e. $LCOE_{ref}$) (compare to Section 4.2), $ASAY_{act}$ the actual average annual electricity yield, annuity of cost the $LCOE_{ref}$ times the reference electricity yield ($ASAY_{ref}$) and B_{env} the newly introduced environmental bonus leading to an increased investment incentive. With costs, yield and FIT rate given the bonus can be determined as a function of a desired ΔRem which however, remains a rather political than purely technical decision. In general it would also be conceivable to introduce a "negative" bonus on those options that are environmentally inferior, consequently reducing the investment incentive but risking the cost and sufficient rate of return covering remuneration. If it is challenged the investment incentive could disappear, leading to a breakdown of the market for the respective technology or installation option. In theory, the support scheme would perfectly represent the environmental performance if the normalized investment incentives (ΔRem values) were equal to the normalized indicator results. Unfortunately a simple back-calculation according to Formula (4.1) in combination with Formula (5.1) is not possible since the different FIT levels granted by the EEG (Tab. 4.2) were assumed to be cost covering only for a specific type and size of installation. Mixing the normalized values among the installation options consequently challenges the assumptions concerning cost recovery made above. What remains possible is the acknowledgment of deviating electricity yields and the magnitude of resulting ΔRem values. Theoretically, cost and equivalent rate of return are covered with a ΔRem equal to zero which would allow to use possible extra profits from diverging electricity yields for a redistribution among the installation options according to environmental criteria. Unfortunately, the electricity yield is location dependent and the amount for redistribution therefore not fixed and thus unsuitable for a nation-wide framework. Reichmuth (2011, pp. 123-137) discussed a model of regional differentiation for the EEG PV

FIT scheme in which the support should be oriented on LCOE as a function of the location dependent insolation. Such a regionally oriented scheme would enable to identify the environmental bonus more easily. But, introducing it cannot be recommended from an economic and environmental efficiency point of view because it would increase the total support cost and, at the same time, lead to the installation of systems with higher specific environmental impacts ($impact/kWh$ or $impact/m^2$) at low yield locations.

Relating the bonus option to the results from the comparison of carbon footprint or EPBT to the EEG support would lead to the conclusion that only slight adaptions were necessary in order to achieve a higher degree of environmentally efficiency in the support scheme. Especially the flat roof (horizontal) installation is undervalued relatively to other options. By for instance granting a bonus of $0.3 ct/kWh$ in prices of 2013 for a flat roof (horizontal) installation the investment incentive would increase to the integrated slanted roof level and thus assure a remuneration close to its environmental performance. Accounting for technological differences (Fig. 4.5) via a bonus would require to reward PV systems deploying a certain module technology that is environmentally superior over others. However, from the comparison above it could be observed that CIGS modules outperformed c-Si modules in terms of carbon footprint and EPBT and CIGS modules already provide higher investment incentives per installed unit of capacity. Hence, a bonus for consolidating the incentive to invest in the environmentally superior technology would not be necessary, but could lead to even higher capacity enlargements with CIGS modules if it were introduced anyway, in turn risking free-rider effects.

When trying to orient the support scheme on the $NEA_{space,LT}$ performance of different installation options more modifications would be necessary (Fig. 4.6). Generally, a relative undervaluation of an installation option by the support scheme is easier to account for than an over-representation, due to the fact that lowering a FIT for a specific installation type could challenge cost recovery and investments in this type. This is for instance the case with flat roof (normal) and open ground systems. Here the investment incentive induced by a high ΔRem over-values the environmental performance of such systems. Accordingly, a negative bonus would lower investments in these installation options. On the other hand, all other options could receive an equivalent positive bonus as a compensation for their environmental superiority increasing their profitability relatively to the environmentally inferior options. Moreover, an additional bonus should then be granted for those installation options that would still remain relatively undervalued, e.g. facade solutions and flat roof (east/west and horizontal). Considering technological differences a bonus option could favor the less space intensive c-Si modules over CIGS modules and consequently increase the efficiency of high yield location deployment.

From the findings of the $NEA_{space,LT}$ it can be concluded that a bonus based on the space intensity of PV installations could be the simplest way of accounting for this environmental criterion. To implement it in practice corridors or limit values for space intensity, defined as the location adjusted electricity yield per m^2 consumed space, could be fixed. Within these corridors a bonus would then be granted depending on the space intensity of the PV installation (similar to the currently applied degression rate corridors in the EEG). Orienting the FIT scheme on space intensity could also be beneficial from a dynamic point of view because producers and investors can account for possible additional earnings and consequently accelerate development of PV system types according to space efficiency criteria.

Leaving aside the installation and technological differences revealed by the three environmental indicators and aiming for a general improvement of the absolute environmental performance of the PV support system, there is the option of applying the bonus on electric efficiency criteria. Even though the analysis in Section 4.2 showed that the actually received electricity yield provides a main constituent of the investment incentive, an additional bonus for deploying a PV system with state of the art PV modules (i. e. a high conversion efficiency) or systems with a high performance ratio[2] would increase the incentive to choose such PV systems over those with lower yields. This option can also account for possibly higher investment costs of the more efficient technology and at the same time improve the overall environmental performance of the installed capacity contributing to dynamic environmental efficiency. Interestingly, it could be shown in Section 2.2.2 that such an option was already under discussion in the research reports of the EEG (cf. Staiß et al. 2007, p. 278) but was refused in order to avoid too much complexity.

Problems of a bonus-based approach arise if the microeconomic choice between different PV technologies is restricted due to technological or other circumstances of installation as described above. In this case windfall gains for the only selectable but particularly promoted options might loom. Furthermore, bonus schemes that only intend to influence comparative technology choice of different PV applications might lead to promote even inefficient (inframarginal) producers or sites that are lifted up to their break-even point.

However, an advantage of the bonus option is the relatively simple way of orienting an established FIT support scheme on environmental criteria and increase its environmental efficiency and thus provide the possibility to influence the behavior of investors and plant operators (cf. Klein et al. 2008, p. 58). Moreover, investment security and the stable investment environment are not negatively affected since the bonus payments are, at least in theory, anticipative. However, it comes at the expense of increasing the complexity of the support scheme (cf. Klein et al. 2008, p. 58) since additional regulations and exceptions have to be introduced, possibly leading to law cross referencing or extensive annexes in the law as it was for instance the case for the biomass boni in the EEG (2009); or the bonus granting would need to be linked with the need of certification procedures entailing additional effort and transaction cost. From an economic point of view the bonus, which grants an additional remuneration above the cost covering FIT level, could lead to a free-rider problem. Investors with a limited ability to choose from different PV in-

---

[2] Consult Tab. 3.2 and Tab. A.2 or Töpfer (2012, pp. 81-83) for an explanation of the terms and underlying values of this study.

stallation options (e.g. because of their available space and roof type) would receive a higher remuneration for a PV system in which they also would have invested without the environmental bonus. Moreover, technological diversity and learning effects could be constrained by the bonus option leading to a possible lock-in development. According to Mendonca et al. (2010, p. 55) it was discussed in Germany to pay a bonus on the FIT for thin-film PV modules but *"the legislator decided to be technology-neutral within this one technology area in order to avoid market interventions at an early stage of technological development. This avoided picking a winner and directing technological development in a certain direction without sufficient information about the future potential of each of these technological options."* This aspect though, could be circumvented when granting boni not on a specific technology or installation option but on more general criteria that indirectly lead to the favoring of certain technologies and installation types. An example would be the above-mentioned orientation of a bonus option on space or electric efficiency. From an economic point of view the environmental bonus would increase the total support cost over what is necessary to achieve economic efficiency (cost recovery and reasonable rate of return). Adding boni on the FIT would therefore raise the EEG differential cost and thus the reallocation charge that has to be borne by the electricity consumers. In the current discussion of overwhelming support cost (especially for PV) and the planned ceiling of the EEG reallocation charge (cf. Altmaier 2013), such an option would probably receive only very limited political acceptance. Total EEG support costs are expected to peak in the mid of this century (cf. UBA 2011, p. 10). Thus, introducing cost increasing measures might face higher social and political acceptance from this point on. On the other hand, the increased environmental efficiency could to some extent also reduce social costs due to higher GHG avoidance and lower external costs.[3] Moreover, higher efficiency could lead to more renewable electricity generation from the same operating capacity and consequently contribute to reach the quantitative EEG goals earlier (e. g. § 1 (2) EEG 2012-PV). A possible compromise between environmental

---

[3] Leaving possible interferences with other policies unaddressed. See Section 5.2.3 for a discussion of interaction policies.

bonus and accruing additional support costs could also be an adapted financing mechanism of the bonus payments, e. g. government financed and borne by the general public instead of being attributed to the EEG reallocation charge and borne by the electricity consumer.

Besides the bonus option, it would also be conceivable to adjust other FIT design parameters (see list in Section 2.1.4). The duration of support could for instance be prolonged for environmental superior options resulting in an increased net present value of the investment for the system operator. The degression rate could also serve as a mean for giving additional incentives to choose the environmentally preferable PV system. Since the degression rate has the twofold purpose of anticipating future learning effects and cost trends but also to partly induce them (cf. Staiß et al. 2007, p. 275), it could be argued that an increased degression rate for environmental inferior PV installation and technology options would trigger technological advancements. Besides, a stepped tariff, granting a higher FIT for environmentally superior technology/installation combinations in the beginning of the remuneration period, would be conceivable to increase the initial investment incentive but limit the accruing costs for society ("front loaded tariff", cf. Couture and Gagnon 2010).

With a maturing PV technology, steadily decreasing support, coming close to electricity spot-market prices, and increasing amounts of renewable electricity, market integration aspects become more and more important (cf. Langniß et al. 2009), as it can be for instance identified in the changing focus of the EEG progress and research reports (cf. Reichmuth 2011, pp. 131-208). The latest EEG versions already implemented direct marketing options and tried a market premium model (see Section 2.2.2; Gawel and Purkus 2013) providing market oriented possibilities for PV electricity remuneration. Thus, it would also be conceivable to switch the fixed FIT system at a certain point to a model focusing on market integration aspects. Langniß et al. (2009) discuss three different possible models from which one is the change to a premium price FIT model (Section 2.1.4) which offers a constant premium above the electricity retail price (cf. Couture and Gagnon 2010). Couture and Gagnon (2010) state that *"the premium can be designed either to reflect the environmental and social attributes of renewable energy, or to approximate RE project costs."*[4] (p. 960). Consequently, given the at least partial cost and rate of return coverage by attainable market prices, a premium could be designed as a fixed absolute or relative amount above the electricity prices steering the development of renewable energies according to environmental goals. PV systems with environmental superior performance, high electric efficiency or preferable space intensity could receive the premium payment leading to increased investment incentives in the desired direction. It should be considered that the market-dependent premium FIT model has severe drawbacks in case of immature markets and infant technologies (cf. Klein et al. 2008, p. 54; Mendonca et al. 2010, pp. 41-42; del Río 2012; Gawel and Purkus 2013). Hence, only if the renewable energy generation technology has reached a considerable degree of competitiveness with conventional electricity generation the market-dependent model could be considered as a possibility to steer further capacity adding according to sustainability criteria.

Madlener and Stagl (2005) argue that *"due to the uncertainty inherent in the assessment of certain criteria, the need for readjustments can be higher than under conventional feed-in tariff systems, which can significantly increase investors' planning uncertainties (and hence the risk premia applied) as well as regulatory and administrative costs."* (p. 162) Therefore, it could be recommended from a purely technical point of view to adapt the FIT scheme according to criteria that are (1) easy to assess and monitor (2) comprise low uncertainties and (3) indirectly lead to positive, maybe more uncertain, environmental effects. I. e. in practice the carbon footprint, EPBT or $NEA_{space,LT}$ should not be the basis for actual changes in the FIT scheme but rather provide meaningful arguments for the necessity of adaptions. Adaptions in turn, might rather be based on components of the assessed environmental indicators entailing improved indicator results. In case of the EPBT and carbon footprint this could be for instance an orienta-

---

[4] RE stands for "Renewable Energies"

tion on electrical PV module and system efficiency and for $NEA_{space,LT}$ an orientation of the FIT scheme on space intensity. Higher electricity yield and space intensities of the operating PV capacity will ultimately lead to an improved environmental performance in term of the assessed indicators. This approach should not be static but rather dynamic, because technological progress and innovation over time will also imply changing environmental impacts. Environmental indicators like EPBT, carbon footprint, etc. and their temporal succession are still beneficial means from a political point of view. The indicators as such and their comparison to the PV FITs could be used as an ecological monitoring tool that enables to discover the need for adaptions. Moreover the results of this paper are useful for communicating the environmental performance of the support scheme to policy-makers and the general public. Hence, a first step for approaching the issue of environmental efficiency and reviewing the compatibility of the EEG with its environmental goals would be their acknowledgment in the ecological analysis and monitoring requested by § 65 EEG (2012-PV) and carried out by the EEG progress and research reports.

5.2.2 Other Support Policy Options

Until now the focus was set on adapting a prevailing FIT scheme according to the identified environmental criteria. The analysis can also be extended to other renewable support schemes which might prove to achieve a certain degree of environmental efficiency more easily. For this reason, the section discusses how quota schemes with TGC, possible tendering options (regulatory, quantity-driven generation based support) and others from Tab. 2.1 can address the raised issue. Finon and Perez (2007) argue that under perfect conditions, i. e. no transaction cost and complete information about marginal cost, the marginal price under a tendering scheme, the equilibrium price of a tradable certificate and the FIT would result in exactly the same level of support. In FIT systems the price is fixed by political negotiation letting the market adjust the amount and in quota systems the desired amount of renewable electricity is fixed politically letting the market define the corresponding price (cf. Finon and Perez 2007; Lipp 2007). Unfortunately, both types of support will produce differing outcomes when accounting for uncertainties and information asymmetries, leading to either an over- or underestimation of the socially optimal amount of renewables (cf. Finon and Perez 2007). Thus, the choice of an appropriate support policy is subject of a lively discussion addressing aspects of investment security, remaining risks, flexibility, cost efficiency and also policy effectiveness. Nevertheless, their specific design elements also offer different options to include additional environmental criteria in renewable capacity expansion efforts.

Firstly, tendering or auction based systems shall be examined in more detail. Haas et al. (2011) define them as systems in which predetermined amounts of capacities are tendered and the competitive bidding selects a winner who is granted a fixed tariff payment or investment support over a certain period. Thereby it is possible to directly control the rents of operators/investors ("pay-as-bid system") (cf. Finon and Perez 2007). Consequently, tendering systems promote the least cost options over others and would entail incentives to develop projects with high electricity yields implying lower LCOE for the investor. To account for a variety of renewable technology options the assigned quota is tendered in specific technology bands (e.g. PV, wind, geothermal capacity, etc.), while the band configuration remains a mere political decision (cf. Madlener and Stagl 2005). Moreover tenders count with the advantage of low information asymmetry in comparison to conventional fixed FIT schemes since the investors participating in the bidding process know their cost structure, which is not necessarily the case if policymakers determine remuneration levels (cf. Becker and Fischer, 2013). However, literature suggests poor success of tendering systems in terms of policy effectiveness, i. e. achieving high renewable deployment, as well as arising support cost (cf. Madlener and Stagl 2005; Ragwitz et al. 2007; Butler and Neuhoff 2008; Mendonca et al. 2010, pp. 175-176; Haas et al. 2011). Mendonca et al. (2010, pp. 175-176) for instance state that tendering systems favor few large investors over a variety smaller ones which could lead to "gaming" the bidding process to exclude competitors. Apart from the inherent advantages and disadvantages of auction-based schemes they comprise the possibility to allocate renewable capacity expansion to the bidder with the lowest cost but also according to predetermined criteria, which could for instance be oriented on the environmental performance or other sustainability aspects. In fact, auctioning systems are frequently used to determine eligibility criteria for participants (cf. e.g. Griffin (2013) in the case of offshore wind energy). Verbruggen and Lauber (2012) state that tendering systems *"are realized by bidders that (best) meet the terms of reference set by public authorities."* (p. 638)

A good example where tendering criteria are designed according to environmental aspects is the French PV support scheme. It subdivides supportable PV projects according to size classes: PV projects smaller than $100kWp$ receive a fixed FIT and for projects with a size between $100kWp$ and $250kWp$ planned to be installed on buildings and those over $250kWp$ for an installation on open ground area a tendering system is introduced (cf. CRE 2013a,b). The FITs for smaller plants are granted for 20 years and comprise additional size and location specific criteria (see e. g. Avril et al. 2012 for details). The tenders in turn are announced and executed by the French Energy Regulation Commission (CRE) in accordance to a detailed functional specification list (cf. CRE 2013a,b). Interestingly, the entity bidding on a certain capacity is legally binded to deploy this capacity after winning the auction (e. g. CRE 2013a, p 3), avoiding the above-mentioned drawback of "gaming" the bidding process (cf. Mendonca et al. 2010, pp. 175-176). Participants have to file in an online form describing the intended PV project. The CRE collects all bids and ranks them according to specific assessment criteria with certain weights. From this ranking the responsible ministers pick the tender winners (cf. e. g. CRE 2013a, p. 5). Projects on buildings between 100 and $250kWp$ are ranked based on their price (66.6% weight) and carbon footprint (in $kgCO_2eq/kWp$) performance (33.3% weight). The carbon footprint can be calculated by the bidder based on reference values presented in the specification document (cf. CRE 2013a, pp. 19-28) or by other entities in the PV supply chain. Additionally the quality and environmental management certification of the PV module producer has to be proven and the bidder has to organize and pay take-back and recycling of the PV systems after their lifetime. Approximately the same regulations apply in the tendering of capacities for large scale open ground systems, besides additional direct environmental impacts (biodiversity, wildlife and habitat concerns, type of land use, landscape integration, etc.) (cf. CRE 2013b, pp. 16-21).

The example indicates the strong focus on environmental admission criteria and therefore an allocation of PV capacity to those plants with an improved environmental or carbon footprint performance. The carbon footprint calculation methodology is based on the CED of different PV module components (feedstock silicon, wafering, cell and module processing, glass, foils as well as thermoplastic parts) in combination with country specific CO_2eq electricity mix factors. CED values can be determined based on actual measurements, publications after 2007 or a reference table attached to the specification list. Thus, it would be valid to apply for the tender using LCA data as for instance developed in Töpfer (2012) and outlined here. From the perspective of a bidder or project developer the data gathering effort seems tremendous which is why typically the PV module producers are requested by investors to provide the environmental data[5] which is then certified by an external institute. Consequently, it is also in the interest of PV equipment producers that potential customers are able to win the tender since it would generate revenues for them. This implies an incentive to carry out environmental analyses of PV products and raise awareness about environmental issues in PV production. Moreover, the PV producers should, at least in theory, not only compete in prices but also in environmental performance leading to dynamic improvements. From the authors' experiences with the French legislation several shortcomings became apparent. Firstly, a high uncertainty and lack of transparency in the reference calculation values could be identified (cf. Annex 4, CRE 2013a, 2013b), i. e. they are not based on a peer reviewed publication leaving relative-

---

[5] Authors' experience from the work at Hanwha Q CELLS.

ly large room for interpretations and risking the incomparability among producers or projects participating in the tender process. Moreover, only a few parts of the PV product are included in the reporting obligation from which some are already outdated in PV module production. BoS equipment is generally not accounted for.

Nevertheless, the French support system shows how different policy options for PV support can be combined and tailored to different investor types prevalent in the PV market. Tendering options are only chosen for larger projects which typically entail different investment structures and possibilities than for instance small private house owners. The latter therefore receive a guaranteed FIT comparable to the EEG legislation. The option of designing environmental eligibility criteria for tender processes thus increases overall environmental efficiency since those projects with better environmental performance are chosen, leading to lower net environmental impacts of the PV operating capacity. The negative economic consequences of the FIT bonus discussed above could be circumvented with the tendering process. However, it could be argued that transaction cost and investment uncertainty increase. The process demands high administrative extra efforts of PV equipment producers, installers, investors/operators and governmental bodies and cannot assure that the effort made to participate is rewarded by winning the auctioned capacity. In order to limit support costs price caps or starting prices can be set in the auctioning procedure (cf. Becker and Fischer 2013). Becker and Fischer (2013) further state that competition is necessarily reduced because of the proposed selection criteria, implying a possible negative impact on achieving the lowest cost option and thus economic efficiency. Explicit local content requirements or hidden regulations to favor local industries are a possible outcome of "designing" selection criteria as well. Moreover, it would be required to introduce a well designed sanctioning system to assure that bidders implement their projects in the way they claimed during the tender.

Given the results obtained in this study eligibility criteria for the capacity auction could be focused on either the carbon footprint, EPBT, $NEA_{space,LT}$ or on all of them. It remains debatable if location and project

specific electricity yield projections should be included in the requirements or if only the "pure" environmental impact should be addressed
as it is currently the case in France. Integrating yield projections has the
advantage of favoring projects with better alignment towards the sun,
installed on favorable insolation conditions and can therefore lead to a
lower overall environmental impact per reference unit (e. g. utilized m^2
space or generated kWh). On the contrary, locations which were princi-
pally suitable for PV remain unconsidered because of unfavorable solar
irradiation, leading to "PV hotspots" in low latitude regions. Focusing
only on the environmental impacts per kWp installed capacity would
disregard important differences in PV installation options as revealed in
Section 3.2.

Next, when applying a TGC quota system for supporting renewables
(see Section 2.1.3) it is also possible to increasingly focus on the environmental performance of renewable energy technologies rather than just
on pure cost-efficiency. Tradable certificates entail the advantage that
investors strive to maximize the yield of their renewable energy plants
in order to (1) increase electricity sales on the market and (2) be awarded with a higher number of salable certificates, consequently increasing
revenues. Therefore, the system *a priori* provides incentives in high yield
installation options and suitable locations. However, what remains unaddressed is the improvement of the absolute environmental impacts (e.
g. level of GHG emissions) or the focus on space intensity among other possible factors. Madlener and Stagl (2005) argue that differentiating
the system according to additional environmental, social and economic
criteria could be accomplished by introducing technology bands, i. e.
separated certificate markets for each renewable technology, or by varying the certificate number issued according to the impacts of a specific
renewable technology. Haas et al. (2011) and Ragwitz et al. (2007 p. 138)
state that in order to avoid the problem of maintaining technological
diversity in the TGC system, certificates could possibly be weighted according to predetermined criteria. Capturing and combining these ideas
for PV in particular would enable to orient the value of certificates on
environmental criteria. Presuming that the scheme is technology band-

ed, i. e. differentiates certificates markets in e. g. PV, wind, biomass, etc., PV installation options or module technologies with especially low GHG emissions, primary energy demand (i. e. CED) or high space efficiency could be awarded on a system basis with a higher certificate weight/value than others. Introducing such an option would improve the environmental efficiency of renewables support, i. e. the quota target would be reached with a lowered environmental impact, but lead to additional complexity and higher transaction costs because of the diversification (cf. Ragwitz et al. 2007, p. 138) and possibly also challenge the system's transparency. It would require establishing clear classification and weighting criteria as well as a scientific back-up of the chosen weights. The latter could be for instance provided by LCA. From an economic point of view, it could be argued that providing incentives to invest in another one but the least cost technology increases social costs of the support system. Higher marginal generation costs, from less competition in the market and possibly higher PV system cost, would lead to a higher marginal certificate price induced by the quota amount (cf. Ragwitz et al. 2007, p. 51). Ragwitz et al. (2007, p. 139-140) present the example of the Wallon region in Belgium where certificates are allocated depending on the CO_2 avoidance of a specific renewable plant in comparison to a reference plant. Unfortunately only direct CO_2 emissions are accounted for, leading to a limited applicability of the instrument to technologies as biomass or landfill gas burning. However, the approach as such could also be transferred to other indicators (e. g. the life cycle GHG emissions, CED, carbon footprint, EPBT or space intensity, etc.) and thus provide a viable option to orient a TGC quota scheme on environmental efficiency.

Besides these main systems for renewable support a variety of complementary alternatives are available in order to provide additional incentives in deploying the more environmental friendly PV electricity generation option. An example of a "supply push" tool (cf. IPCC 2012, p. 932) are R&D expenditures which aim at fostering technological progress and innovation or "explorative learning" instead of "exploitative learning" (cf. Hoppmann et al. 2013). Contrary, deployment support policies, as discussed above, are "demand pull" measures which are in general suitable for promoting the market penetration of technologies beyond the demonstration stage (cf. IPCC 2012, p. 932), thus increasingly contributing to exploitative technological learning risking a lock-in among the already established technologies (cf. Hoppmann et al. 2013). Hoppmann et al. (2013) argue that *"such a lock-in may be uncritical in the case that it is certain that these technologies meet present and future needs in terms of economic, ecological and societal dimension"* (p. 1000). Given the inherent uncertainty of future development and needs, technological diversity is seen as desirable and deployment policies should consequently promote a portfolio of technology options (e. g. via diversified FITs, see above) or be accompanied by R&D support (cf. Hoppmann et al. 2013). The 2011 public PV R&D budget was approximately 59 million€ granted by the BMU and the German Federal Ministry of Education and Research (BMBF) supporting c-Si, thin-film, BoS, organic PV cell and concentrating solar technology as well as research in recycling and ecological impact issues of PV systems (cf. Wissing 2012, p. 18). An increased budget for research in fostering the environmental performance of PV systems can therefore also indirectly contribute to environmental efficiency of the operating PV capacity.

Moreover, investment subsidies, tax credits and soft loans as well as tendering systems for investment grants are considered additional measures to promote renewables (cf. Haas et al. 2011). All of them could be granted according to specified environmental criteria. Investment subsidies could favor the deployment of a certain PV installation type or module technology over others and thus provide additional incentives for investing in these PV systems since the investor's rate of return would be increased without any alteration of the remuneration scheme. Basically the same is applicable to tax credits for a certain PV plant operator that can be either based on the investment amount or the amount of electricity generated (cf. Mendonca et al. 2010, pp. 170-173). Soft loans improve the investment conditions as well and could consequently be linked to environmental eligibility criteria. In Germany the reconstruction loan corporation (*Kreditanstalt für Wiederaufbau* - KfW) offers diverse soft loans for renewable energy projects (cf. KfW n.d.). Concerning tenders for investment grants, generally the same selection criteria for participation and bidding could apply as discussed above for the renewable quota tendering system.

A drawback of all options are the necessarily increased support cost, either borne by the public budget, i. e. the tax payer or the electricity consumer as well as conceivable undesirable lock-ins when selection criteria are unbalanced or biased towards a specific technology option which proves to be inferior to others in the future. Hence, it could be recommended to carry out separate cost-benefit analyses in case support schemes are envisaged to be adapted according to environmental criteria or additional support options are introduced that aim to foster environmental efficient installation of PV or renewable capacity in general. Thereby it is of particular importance to account for possible side effects and interactions with other policy instruments as well as additional benefits associated with the generation of renewable electricity. These aspects are discussed in more detail in the upcoming subsection.

5.2.3 Interaction Policies

Literature suggests that technology policies promoting renewables are only a second-best alternative for tackling multiple pollution problems such as climate change and environmental harm (see Section 2.1.1; Lehmann and Gawel 2013; IPCC 2012, p. 872). First-best pollution control policies would internalize externalities by either a price (e. g. Pigouvian Tax) or quantity (e. g. tradable permit scheme) based intervention (cf. Lehmann 2012). Currently, the EU and its member states apply a policy mix rather than a single policy for addressing climate change, including the EU ETS and renewable support schemes (e. g. EEG), among others (cf. Lehmann 2012). The interaction between those two policy tools is actively discussed in literature. Critics argue that the instruments like the EEG cannot contribute anything to climate change mitigation in the prevalence of the EU ETS. The interaction policy argument is based on the following rationale: the increased renewable electricity generation reduces the output of conventional carbon intensive power plants leading to absolutely lower CO_2 emissions in the electricity sector. Thereby the demand for emission allowances is reduced implying a lower price which in turn allows other sectors participating in the ETS to compensate for this effect by increasing their emissions (cf. Gawel and Lehmann 2013; Frondel et al. 2010; Sorrell and Sijm 2003). The result is a shift of emissions from one ETS sector to another. Reaching the fixed emission cap would then come at increased overall cost since cheap emission abatement options are crowded out by more expensive support for renewables (cf. Frondel et al. 2010). Because of its very high emission abatement costs (from 700 to 1000€/kWh) especially the support for PV is challenged by this interaction argument (cf. Lehmann and Gawel 2013).

With these deliberations in mind, increasing the environmental efficiency of the support scheme for renewables or PV in particular seems questionable if not obsolete with regards to climate change or in any case where other pollution problems are internalized by a tradable permit scheme. If the support provided incentives for higher electric and space efficiency more renewable electricity from a given operating capacity would be generated, increasing the CO_2 emission avoidance and probably also total support costs (e. g. because of introduced boni). Because of this, ETS certificate prices would further decline, leading to a shift of more emissions away from the electricity into other ETS sectors and a subsequent increase of the induced welfare loss. Interestingly, it is suggested that Pigouvian Taxes, as a possible first-best policy, entail equivalent internalization possibilities for pollution problems but, in comparison to tradable permits, do not counteract the mitigation effects of simultaneous technology support policies (cf. IPCC 2012, p. 917; Philibert 2011, p. 16).

However, Section 2.1.1 could already show that a support of renewable technologies (EEG) can also be justified in the presence of a specific pollution control policy (EU ETS) by a multiplicity of pursued aims and externalities respectively (e. g. energy security, access to electricity, avoidance of other detrimental environmental impacts, etc.) or policy failures and possible lock-in risks occurring which cannot be overcome with the current EU ETS (Gawel et al. 2013b). Lehmann and Gawel (2013) state that renewable promotion schemes indirectly reduce the emission cap or quota of an ETS because the increased renewable electricity generation is anticipated in the political negotiation process to set this cap. Even more, for reasons of political feasibility, support schemes turn out to be necessary to ease the cost pressure for powerful interest groups in the GHG emission sectors entailed by the EU ETS (Gawel et al. 2013b). Moreover, imperfect competition in the electricity market, subsidies for non-renewable electricity generation, investment uncertainties or market entry and financing barriers, etc., can call for renewable promotion schemes to reduce these barriers and foster the establishment of equal conditions (cf. Lehmann and Gawel 2013; IPCC 2012, pp. 880-882).

Consequently, an increased GHG emission avoidance within renewable promotion scheme, facilitated by measures proposed above, might be, in the presence of the mentioned market and policy failures, not completely offset in combination with an ETS but able to contribute to tighter emission caps and consequently lower absolute GHG emission. Lehmann and Gawel (2013) state that *"by offering a subsidy, the government facilitates the attainment of an ambitious emissions target and thereby 'buys' the agreement of stakeholders which have to reduce their emissions."* (p. 601) Moreover, it should be considered that the proposed support adaption measures entail important side effects as well as possible emission reductions in sectors not participating in an ETS (e. g. in the PV supply chain). Additionally promoting electric and space efficiency can decrease the specific impact of all environmental problems associated with renewable electricity generation (e. g. climate change, water use, land use, heavy metal emissions, air pollution or human health and ecosystem damages in general, etc., see e. g. Fthenakis and Kim (2011) or Turney and Fthenakis (2011) for PV) as well as increasingly avoid indirect environmental impacts associated with conventional electricity pollution (e. g. air pollution and related human health impacts, resource extraction and related eco-system damages, material damages, agricultural losses, accidents, geo-political effects, nuclear waste storage, etc.

see IPCC 2012, p. 855). Avoiding such negative environmental impacts can therefore further reduce corresponding external costs and consolidate positive effects of renewable energies (cf. IPCC 2012, pp. 854-857). However, ascending support costs from the implementation of environmental efficiency increasing measures, e. g. by granting boni for electric efficiency or space intensity of PV systems in the EEG, should certainly be subject to a separate efficiency analysis, comparing accruing costs and benefits to those of other emission mitigation options.

Another argument generally in favor of renewable promotion schemes are possible technological lock-in effects in carbon intensive structures when not introducing them and the need of a long-term perspective. Referring to the latter Philibert (2011, p. 10) states that in the short-term renewable promotion might increase the costs of achieving CO_2 targets but the resulting broad portfolio of clean technology options in combination with expected significant cost reductions (technological learning, see also Section 2.1.1) will be a precondition to reach the outlined long-terms goals of mitigating climate change and entail low future emission abatement costs. Philibert (2011, p. 16) further argues that, given a situation where a ETS is in place and no renewable promotion policy is pursued, the short-term advantage of less carbon-intensive technologies could lead to a lock-in in such technologies. According to Lehmann and Gawel (2013) path dependencies occur in the electricity sector because of increasing returns of technology adaption (e. g. economies of scale, learning economies, adaptive expectations, network economies), large-scale and long-term investments, low product differentiation possibilities and inert institutional change, leading to a system which might not be able to implement changes (i. e. emission reductions) in the required time horizon. A lock-in development in carbon-intensive pathways could consequently bear *"the risk that the cost of switching to alternative technologies could become prohibitive, particularly if climate impacts turn out to be more severe than anticipated."* (Sorrell and Sijm 2003, p. 430) Therefore, early targeted support in areas with low-carbon innovation opportunities is recommended (cf. Sorrell and Sijm 2003). This argument could be a conceivable justification for additionally supporting the climate-friendliness of the installed renewable capacity, even though the support might imply increased short-term costs.

Ultimately, the merit-order effect might be fostered with the adaption of a support scheme towards higher amounts of renewable electricity generation (e. g. from FIT boni for space efficiency or electric efficiency). With increasing preferably purchased renewable electricity, entailing low marginal generation costs, more expensive electricity from conventional sources is crowded out implying a decrease of spot-market electricity prices (cf. Mennel 2012; Sensfuß 2011b p. 3). Sensfuß (2011b, p. 9) estimates economic savings in the year 2010 induced by the merit-order effect of approximately 2.8 billion€. From this point of view, merit-order effects could justify additional support for environmental efficiency, i. e. electricity yield increasing measures. However, more complex interrelations lead to an unfortunate paradox. With decreasing spot-market prices and possibly increasing support costs, the differential costs of the support scheme (support costs minus revenues from electricity sale) rise, increasing the burden on society. Moreover, the decrease in prices for CO_2 allowances from the EU ETS in combination with renewable promotion schemes (see above) aggravates the increase in differential costs since it further decreases the electricity price. In fact, decreased spot-market electricity prices contribute with 16% to the total 2013 EEG reallocation charge and therefore burden the electricity consumer in the short-term (BEE 2012). Hence, such interacting effects should be accounted for when trying to quantify the possible benefits and costs of an environmentally optimized renewable or PV support scheme.

ronmentally inefficient operating PV capacity, Chapter 3 developed a set of exemplary indicators from an LCA of eight different PV installation options and Mono-Si, Multi-Si and CIGS module technologies. The indicators exemplary selected were the carbon footprint, $NEA_{space,LT}$ and EPBT. They were intended to represent criteria reflecting the EEG's environmental goals (see Section 3.1). Additionally, it could be identified that the selection process is not only a technical but also a political and normative decision and consequently depends on aspects of "what can be measured" as well as of "what should be measured" (cf. McCool and Stankey 2004). Therefore, scientific indicator quality criteria as well as political goals and their prioritization play a role for the development of appropriate indicators. Out of this insights and in case of an advancement of the ideas presented in this survey it was proposed to extent the indicators assessed to a broader set, accounting for a more diverse environmental impacts, while maintaining easy communicability. As an example, UNEP (2010, p. 37) provides a conceivable list of impacts and indicators to measure and communicate them. Moreover, it would be possible to expand the analysis' focus away from the environmental to a holistic life cycle sustainability assessment, acknowledging social and economic aspects derived from the EEG objectives (listed in Tab. 2.2).

The comparison of the FIT scheme, or more precisely, the investment incentives it provides, to the carbon footprint and EPBT of the 24 PV installation/technologies combinations considered in the LCA revealed a good coincidental match (Section 4.3) in all assessed EEG law versions. Thus, from this point of view the EEG versions would contribute to the climate and resource efficient installation of PV capacity. The findings were obtained independently for different installation types and technologies and are mainly attributable to the fact that both, indicators and remuneration from the FIT scheme, are dependent on the amount of electricity generated by PV systems. I. e. higher electricity yields improve the environmental indicator of and simultaneously provide higher investment incentives in a certain PV system. Conversely, this match could not be confirmed in case of the $NEA_{space,LT}$ (Section 4.4). Concerning this indicator it could be concluded that the PV FITs provide rather opposing investment incentives since their introduction in 2000. CIGS technology for instance counts with a low $NEA_{space,LT}$ performance because of its relatively low electric efficiency but is preferred over the superior c-Si technology under the prevailing support conditions. In terms of installation options, especially flat roof (normal) and open ground PV systems were found to provide relatively too high investment incentives in comparison to their $NEA_{space,LT}$ performance. The comparison results are necessarily based on the assumption that FITs granted by the EEG are exactly cost and rate of return covering under the reference conditions they were developed. Deviations from these conditions, e. g. by differing electricity yields of PV systems receiving an equivalent FIT, were supposed to constitute the investment incentive in such PV systems. An increased investment incentive would then facilitate capacity enlargements of such a PV system on a macro scale and entail certain environmental effects of the installed and EEG remunerated PV capacity (Section 4.2).

Hence, this contradiction in the obtained comparison results led to the conclusion that sub-research question (2) cannot simply answered with yes or no. Therefore, Section 5.1 pointed at the importance of choosing appropriate environmental indicators with specific reference units and presented MCDA as an approach to cope with contradictory indicator results and rising complexity when considering more assessment criteria, in order to derive meaningful policy implications and a transparent goal prioritization.

Ultimately, a multitude of conceivable options to adapt the EEG FIT scheme for PV according to the revealed mismatches between investment incentive and environmental performance were identified (Section 5.2.1). Granting an environmental bonus on-top of the prevailing FIT level was found to be a relatively simple way to steer investment incentives in a desirable direction without challenging the investment security of the system. On the other hand, implementing such an option would conceivably trigger free-rider effects and increase the complexity of the scheme as well as the total support cost borne by the public or the electricity consumer, possibly receiving limited acceptance in the current EEG cost containment debate. Furthermore, different other design options were discovered as being suitable to reflect the environmental performance of the assessed PV installation options and technologies in the EEG. Among them are adaptions of the support duration, degression rate or the introduction of a front loaded FIT. Once electricity generation from PV has reached competitiveness it would also be conceivable to switch the currently applied fixed FIT scheme to a premium price model where support granted is completely linked to the environmental or sustainability performance of a PV system type. In summary, it could be recommended to adapt the prevailing EEG PV FIT scheme according to criteria that are (1) easy to assess and monitor (2) comprise low uncertainties and (3) indirectly lead to positive, maybe more uncertain, environmental effects. Precisely, directing additional support towards high electric or space efficiency instead of basing it on detailed environmental indicators could increase the environmental efficiency of the newly installed PV capacity, lower the environmental impacts in absolute terms and would favor particularly efficient technologies and installation options over others. Simultaneously, a "picking a winner" in terms of PV technologies or installation options would be avoided.

The discussion revealed that not only the FIT scheme as applied by the EEG can be subject to adaptions according to environmental criteria but also other renewable deployment support policies (e. g. quota tendering or TGC) and support options (e. g. public R&D expenditures, see Section 5.2.2). In either case, technology-specific support is still needed in the future. Amongst others, the rationale here is for reasons of heterogeneous environmental impacts. Approaches that deny heterogeneity of electricity generating technologies (such as simple overall quota mechanisms) are not able to incorporate different ecological impacts of the respective technologies. This way, the future technology mix is just not irrelevant from an environmental economics point of view (Gawel et al. 2012a), but anyhow repeatedly suggested by scholars invoking a more general economics argument (see e. g. Frondel et al. 2008, 2010).

However, all adjustments would bring along either directly or indirectly increased support costs and might entail technology lock-ins if design criteria are biased and unbalanced. These drawbacks must essentially be traded off against the positive environmental effects and conceivable external cost savings. Hence, carrying out cost-benefit analyses to evaluate the introduction of environmental efficiency improving components in renewable support policies could be a topic for further research and seems fruitful in case such adaptions were envisaged. These deliberations should also encompass positive and negative side effects evoked by interacting policies. As discussed in Section 5.2.3, interferences with the EU ETS, possibly leading to a partial CO_2 emission reduction offset, and the paradox of merit-order effects would require attention in such assessments.

In summary, it can be inferred that the PV FIT design in the EEG is only partially accounting for criteria which reflect the main environmentally oriented goals of the law. The issue of environmental efficiency in the PV capacity expansion receives rather low attention in the monitoring process and consequently in the law's implementation. Out of this, the study presented an exemplary environmental efficiency assessment of the EEG PV FIT scheme and investigated a set of adaption options for the prevailing PV support as well as other conceivable renewable promotion policies. Three main recommendation can be derived from the analysis: firstly, to address the issue of environmental efficient goal achievement in regular monitoring processes (e. g. in the EEG progress and research reports). This dynamic procedure could be based on actually observed PV capacity expansions in combination with the presented assessment approach. Secondly, to identify the necessity adapting the support scheme from the environmental efficiency assessment. And thirdly, in case adaptions are required, to change the support scheme based on criteria which are transparently developed, comprise low uncertainty, are easy to assess and, at the same time, entail the envisaged positive environmental performance. As an example, this paper proposed to provide additional investment incentives in space efficiency and electric efficiency. Hence, the presented approach can be beneficial for both, the identification of environmental inefficient incentives provided by support schemes and the derivation of suitable adaption possibilities to cope with these inefficiencies.

The approach chosen here to assess the EEG PV FIT scheme can certainly be challenged. Investment incentives could for instance be determined more precisely when specific cost data for reference PV installation types is available. Thereby the constraining assumptions of Section 4.2 could be, at least partially, relaxed. Moreover, using peer-reviewed environmental LCA data and studies instead of data from a two specific producers, could contribute to the robustness, representativeness and validity of the obtained indicator results. As the sensitivity analysis has shown changing basic assumption in the environmental model can significantly alter the indicator performance and challenge the applicability of the model on macro scale (Section 3.4). Hence, basing the environmental indicators on transparently modeled average and up-to-date PV industry data could be fruitful for a future research and monitoring. The presented approach is also expandable to a holistic sustainability assessment of support schemes including social and economic aspects or to other renewable electricity generation technologies (e. g. electricity from water, wind, biomass, etc.). The latter could be of particular interest because it would enable to investigate deviations in EEG remuneration of the promoted renewable technologies in comparison to the impacts associated with their electricity generation. With such an analysis the overall environmental efficiency of the EEG and consequently its effectiveness in goal achievement (and thus its efficiency with respect to the full set of targets including environmental impacts) could be determined.

Annex

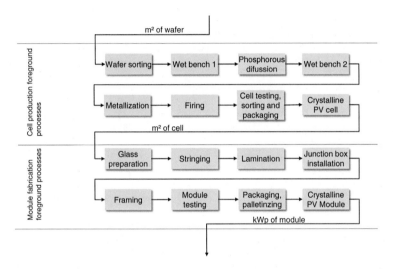

Fig. A.1: Main processes of c-Si PV cell and module production with corresponding reference flows considered in LCI modeling; Source: Authors

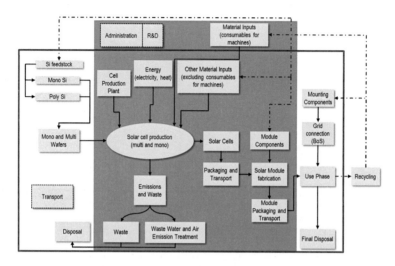

Fig. A.2: LCA System boundaries of c-Si PV systems; Source: Authors

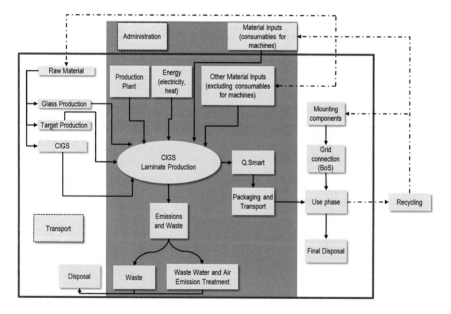

Fig. A.3: LCA System boundaries of CIGS PV systems; Source: Authors

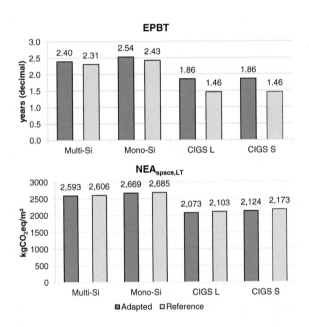

Fig. A.4: Effects of foreground processes using the German electricity mix on the EPBT and $NEA_{space,LT}$; Source: Authors

Tab. A.1: Product systems considered in the assessment; Source: Authors

Installation type	Examplary pictures for mounting	Module Inclination	Data Sources
Slanted Roof (mounted) (residential scale)	mounted on roof	Optimal (35°)	Company specific cells and modules Ecolnvent premodeled processes for up- and downstream processes (including silicon, wafer and BoS components)
Slanted Roof (integrated) (residential scale)	integrated in roof, crystalline with frame	Optimal (35°)	Company specific cells and modules Ecolnvent premodeled processes for up- and downstream processes (including silicon, wafer and BoS components)
Flat Roof (normal) (commercial scale)	rack on flat roof, one module	Optimal (35°)	Company specific cells and modules Ecolnvent premodeled processes for up- and downstream processes (including silicon, wafer and BoS components)
Flat Roof (east/west) (commercial/industrial scale)	rack on flat roof, triangle	10°	Company specific cells, modules and mounting and cabling structure. Ecolnvent premodeled processes for up- and downstream processes (including silicon, wafer and inverter components)
Flat Roof (horizontal) (commercial/industrial scale)	rack on flat roof, horizontal mounting	3°	Company specific cells, modules and mounting and cabling structure. Ecolnvent premodeled processes for up- and downstream processes (including silicon, wafer and inverter components)
Facade (mounted) (residential/commercial scale)	mounted on facade	90°	Company specific cells and modules Ecolnvent premodeled processes for up- and downstream processes (including silicon, wafer and BoS components)
Facade (integrated) (residential/commercial scale)	integrated in facade, crystalline with frame	90°	Company specific cells and modules Ecolnvent premodeled processes for up- and downstream processes (including silicon, wafer and BoS components)
Open Ground (Utility scale)	rack mounted on open ground, 2 (crystalline) or 4 (CIGS)modules one upon the other	Optimal (35°)	Company specific cells and modules Ecolnvent premodeled processes for up- and downstream processes (including silicon, wafer and BoS components)

Tab. A.2: LCA transparency reporting parameters; Source: Authors, according to IEAPVPS 2011

Reporting to comply with the IEAPVPS (2011) methodology guidelines on PV LCAs:
1. On-plane irradiation level and location
Reference Location: Munich (48°8'11" North, 11°34'37" East, Elevation 521m)
Irradiation: 1,142 kWh/m²/year (calculation based on PVGIS-classic database of the European Commission Joint Research Center (ECJRC n.d.)
2. Module-rated efficiency
Mono-Si: 15.3% (Q.Peak 255Wp module)
Multi-Si: 14.7% (Q.Pro 245Wp module)
CIGS L: 11.2% (Q.Smart L 105Wp module)
CIGS S: 11.4% (Q.Smart UF 85Wp module)

3. System performance ratio (at reference location)

Performance ratio	Multi-Si	Mono-Si	CIGS L	CIGS UF
Slanted roof (mounted)	81%	80%	84%	85%
Slanted roof (integrated)	78%	77%	84%	82%
Flat roof (normal)	83%	83%	85%	86%
Flat roof (east/west)	79%	78%	81%	82%
Flat roof (horizontal)	82%	81%	84%	85%
Facade (mounted)	79%	79%	81%	82%
Facade (integrated)	76%	76%	79%	80%
Open ground	85%	85%	89%	89%

4. Time frame of data
Time of data collection: 2011 for c-Si and 2010 for CIGS
5. Type of system
Eight grid connected fixed tilt installation options:
–   Slanted rooftop installation: mounted and integrated
–   Flat roof (normal): Slant module alignment in subsequent rows
–   Flat roof (east/west): Tent like module installation facing east/westwards
–   Flat roof (horizontal): 3° tilted
–   Façade installation: mounted and integrated
–   Open ground: mounted fixed tilt system
6. Expected lifetime for PV and BoS
Lifetime of modules, structure and cabling assumed: 30 years
Lifetime of inverter: 15 years
7. System Boundaries
Refer to Fig. A.2 and Fig. A.3
8. Place/Region of production
LCI model according to Hanwha Q CELLS and Solibro supply chain with c-Si wafer production in Europe and Asia, c-Si cell production Germany and c-Si module production in Europe and Asia. CIGS module production is located in Germany. BoS parts are based on specific EcoInvent 2.2 processes (commonly Europe). Electricity mixes refer to the corresponding region/country of production.
9. Goal of the study
Environmental impact assessment (EPBT and carbon footprint) of eight different grid connected PV installation possibilities for the use in a comparative analysis of the German PV FIT's. The LCA is classified as retrospective with large scale decision support (cf. ECJRC 2010) based on current performance technology (cf. IEAPVPS 2011). The commissioning entity is the University of Leipzig in cooperation with Hanwha Q CELLS.
10. Degradation ratio assumed
Ratio: 0.6%/year
11. LCA method used
Process based
12. LCA tool and database used
SimaPro 7.3 in combination with Ecolnvent 2.2
13. Assumptions for production of major input materials
Background processes are based on EcoInvent 2.2.
The electricity source in foreground processes was adapted to the actual mix used by the producer. Silicon background processes have been adapted to the place of production (electricity mix) and to the wafer thickness used in the cell production process of Q CELLS. A mix of primary and secondary production of aluminum was chosen from the EcoInvent database.

Tab. A.3: ASAY of different technology/installation combinations at the reference location Munich; Source: Authors

ASAY (kWh/kWp)	Inclination	Multi-Si	Mono-Si	CIGS
Slanted Roof (mounted)	35°	972.5	967.9	1,011.9
Slanted Roof (integrated)	35°	933.0	927.5	980.7
Flat Roof (normal)	35°	998.1	994.5	1,033.0
Flat Roof (east/west)	10°	825.4	820.9	856.3
Flat Roof (horizontal)	3°	921.1	916.7	952.3
Facade (mounted)	90°	669.9	666.9	690.6
Facade (integrated)	90°	647.1	643.6	672.0
Open Ground	35°	1,023.8	1,022.9	1,072.0

Tab. A.4: Space requirements for different PV installation systems; Source: Authors

m² space occupied per kWp	Multi-Si	Mono-Si	CIGS L	CIGS S
Slanted Roof (mounted)	6.82	6.55	8.94	8.82
Slanted Roof (integrated)	6.82	6.55	8.94	8.82
Flat Roof (normal)	17.25	16.57	22.67	22.40
Flat Roof (east/west)	6.71	6.45	8.78	8.68
Flat Roof (horizontal)	6.82	6.55	8.94	8.82
Facade (mounted)	6.82	6.55	8.94	8.82
Facade (integrated)	6.82	6.55	8.94	8.82
Open Ground	17.27	16.59	22.70	22.33

Tab. A.5: ASAY per m^2 occupied space for different PV installation systems at Munich; Source: Authors

Space related ASAY (kWh/m² occupied space)	Multi-Si	Mono-Si	CIGS L	CIGS S
Slanted Roof (mounted)	142.67	147.79	112.55	115.35
Slanted Roof (integrated)	136.88	141.62	109.06	111.82
Flat Roof (normal)	57.88	60.02	45.33	46.36
Flat Roof (east/west)	122.93	127.26	96.93	99.22
Flat Roof (horizontal)	135.13	139.97	105.88	108.59
Facade (mounted)	98.28	101.83	76.75	78.78
Facade (integrated)	94.94	98.27	74.70	76.66
Open Ground	59.30	61.66	47.13	48.11

Tab. A.6: Effects of electricity consumption in the crystallization phase
on the indicator results of c-Si PV systems; Source: Authors

Mono-Si	EPBT (years)	%-change	Carbon Footprint (gCO₂eq/kWh)	%-change	NEA$_{space,LT}$ (kgCO₂/m²)	%-change
Slanted Roof (mounted)	2.20	-9.31%	46.37	-9.78%	2707.32	0.83%
Slanted Roof (integrated)	2.27	-9.43%	47.68	-9.91%	2588.85	0.87%
Flat Roof (normal)	2.15	-9.28%	45.47	-9.71%	1101.12	0.81%
Flat Roof (east/west)	2.50	-9.66%	51.63	-10.30%	2311.10	0.99%
Flat Roof (horizontal)	2.27	-9.53%	47.27	-10.09%	2560.35	0.88%
Facade (mounted)	3.16	-9.41%	66.52	-14.42%	1803.79	1.25%
Facade (integrated)	3.33	-9.27%	70.18	-9.72%	1729.99	1.30%
Open Ground	2.27	-8.61%	49.35	-8.79%	1124.11	0.79%

Multi-Si	EPBT (years)	%-change	Carbon Footprint (gCO₂eq/kWh)	%-change	NEA$_{space,LT}$ (kgCO₂/m²)	%-change
Slanted Roof (mounted)	2.25	-2.78%	46.69	-2.94%	2612.08	0.23%
Slanted Roof (integrated)	2.31	-2.82%	47.94	-2.98%	2501.02	0.24%
Flat Roof (normal)	2.20	-2.77%	45.85	-2.92%	1061.18	0.23%
Flat Roof (east/west)	2.55	-2.89%	51.95	-3.11%	2231.45	0.28%
Flat Roof (horizontal)	2.32	-2.85%	47.62	-3.04%	2470.33	0.25%
Facade (mounted)	3.23	-2.82%	66.98	-7.92%	1739.56	0.35%
Facade (integrated)	3.40	-2.77%	70.62	-2.92%	1670.11	0.36%
Open Ground	2.34	-2.55%	50.23	-2.60%	1079.44	0.22%

Tab. A.7: Effects of foreground processes using the German electricity mix on the EPBT, Carbon Footprint and $NEA_{space,LT}$ (complete results table); Source: Authors

EPBT (years)	Multi-Si adapted	%-change	Mono-Si adapted	%-change	CIGS L adapted	%-change	CIGS S adapted	%-change
Slanted Roof (mounted)	2.40	3.76%	2.54	4.29%	1.86	27.01%	1.86	27.59%
Slanted Roof (integrated)	2.47	3.81%	2.61	4.35%	1.88	26.70%	1.89	29.29%
Flat Roof (normal)	2.35	3.74%	2.48	4.28%	1.83	25.79%	1.84	28.43%
Flat Roof (east/west)	2.72	3.90%	2.89	4.45%	2.09	27.68%	2.10	30.51%
Flat Roof (horizontal)	2.48	3.85%	2.62	4.39%	1.92	26.87%	1.93	29.63%
Facade (mounted)	3.45	3.80%	3.64	4.34%	2.68	27.59%	2.68	28.25%
Facade (integrated)	3.62	3.74%	3.83	4.28%	2.82	26.78%	2.82	27.32%
Open Ground	2.48	3.44%	2.58	3.97%	2.06	21.20%	2.07	23.43%

Carbon Footprint (gCO₂eq/kWh)	Multi-Si adapted	%-change	Mono-Si adapted	%-change	CIGS L adapted	%-change	CIGS S adapted	%-change
Slanted Roof (mounted)	51.05	6.12%	54.93	6.88%	43.09	43.22%	43.09	48.44%
Slanted Roof (integrated)	52.48	6.21%	56.61	6.98%	43.54	44.54%	43.71	49.67%
Flat Roof (normal)	50.10	6.08%	53.81	6.84%	42.66	42.56%	42.66	47.69%
Flat Roof (east/west)	57.08	6.47%	61.73	7.25%	47.63	47.65%	47.65	53.51%
Flat Roof (horizontal)	52.22	6.33%	56.31	7.11%	44.19	45.51%	44.18	51.07%
Facade (mounted)	73.31	6.19%	78.95	6.96%	62.15	44.25%	62.09	49.60%
Facade (integrated)	77.17	6.08%	83.06	6.85%	65.50	42.67%	65.40	47.82%
Open Ground	54.37	5.43%	57.46	6.19%	49.21	33.08%	49.26	37.04%

NEA_space,LT (kgCO₂eq/m²)	Multi-Si adapted	%-change	Mono-Si adapted	%-change	CIGS L adapted	%-change	CIGS S adapted	%-change
Slanted Roof (mounted)	2593.43	-0.48%	2669.35	-0.58%	2059.86	-2.07%	2124.48	-2.24%
Slanted Roof (integrated)	2482.37	-0.51%	2550.88	-0.61%	1994.53	-2.14%	2057.26	-2.31%
Flat Roof (normal)	1053.81	-0.47%	1086.11	-0.57%	835.13	-2.03%	854.46	-2.19%
Flat Roof (east/west)	2212.51	-0.58%	2272.54	-0.70%	1757.09	-2.46%	1813.86	-2.65%
Flat Roof (horizontal)	2451.67	-0.51%	2522.38	-0.62%	1934.33	-2.21%	1996.41	-2.38%
Facade (mounted)	1720.91	-0.73%	1765.82	-0.88%	1361.09	-3.11%	1406.02	-3.34%
Facade (integrated)	1651.46	-0.76%	1692.02	-0.92%	1317.26	-3.21%	1360.54	-3.45%
Open Ground	1072.08	-0.46%	1109.12	-0.56%	859.42	-1.97%	877.14	-2.14%

Tab. A.8: Sensitivity analysis location characteristics; Source: Authors, based on references cited in the table

	Optimal module tilt	Annual irradiation on horizontal (kWh/m²/year)	Rprim (MJ$_{prim}$/kWh$_{end}$)	CO$_2$eq emissions (gCO$_2$eq/kWh)
Munich	35°	1,142	11.6	657
Cadiz	30°	1,745	10.6	526
Toulouse	35°	1,361	12.3	92
Las Vegas	27°	2,285*	12.9	775
London	34°	986	10.9	617
Rome	30°	1,471	9.7	650
Cape Town	27°	2,037	11.4**	531**
Source	PV Sol Expert 4.5 (R1)	ECJRC (n.d.)	Ecoinvent 2.2 with CED 1.08 method	Ecoinvent 2.2 with IPCC 2007 GWP 100a method

*Source: NREL (n.d.)
** Due to a lack of data the electricity mix of the Union for the Co-ordination of Transmission of Electricity (UCTE) has been used according to recommendations of Frischknecht et al. (2007, p. 11)

Tab. A.9: EPBT and carbon footprint indicator results for Cadiz, Spain; Source: Authors

EPBT (years) – Cadiz	Multi-Si	Mono-Si	CIGS
Slanted Roof (mounted)	1.65	1.73	1.08
Slanted Roof (integrated)	1.73	1.81	1.11
Flat Roof (normal)	1.62	1.69	1.08
Flat Roof (east/west)	1.83	1.92	1.18
Flat Roof (horizontal)	1.79	1.88	1.18
Facade (mounted)	2.76	2.88	1.82
Facade (integrated)	2.94	3.07	1.95
Open Ground	1.76	1.82	1.31

Carbon Footprint (gCO$_2$eq/kWh) – Cadiz	Multi-Si	Mono-Si	CIGS
Slanted Roof (mounted)	31.48	33.51	20.44
Slanted Roof (integrated)	32.93	35.15	20.81
Flat Roof (normal)	31.05	33.01	20.40
Flat Roof (east/west)	34.39	36.75	21.32
Flat Roof (horizontal)	33.89	36.15	21.70
Facade (mounted)	52.73	56.01	34.40
Facade (integrated)	56.21	59.66	37.02
Open Ground	34.77	36.48	26.24

Tab. A.10: Normalized carbon footprint, EPBT and $NEA_{space,LT}$ indicator results; Source: Authors

Carbon Footprint	Multi-Si	Mono-Si	CIGS L	CIGS S
Slanted Roof (mounted)	60.65	53.92	97.54	99.70
Slanted Roof (integrated)	57.98	50.79	97.45	99.34
Flat Roof (normal)	62.45	56.03	97.87	100.00
Flat Roof (east/west)	49.37	41.30	93.09	95.58
Flat Roof (horizontal)	58.59	51.50	96.96	99.26
Facade (mounted)	17.81	8.02	70.92	74.16
Facade (integrated)	10.21	0.00	65.15	68.55
Open Ground	53.56	48.36	83.43	85.54
$NEA_{space,LT}$	Multi-Si	Mono-Si	CIGS L	CIGS S
Slanted Roof (mounted)	95.69	100.00	68.98	72.06
Slanted Roof (integrated)	89.63	93.53	65.40	68.39
Flat Roof (normal)	11.24	13.07	0.00	1.14
Flat Roof (east/west)	74.91	78.36	52.61	55.15
Flat Roof (horizontal)	87.95	91.98	62.09	65.07
Facade (mounted)	48.07	50.69	30.61	32.85
Facade (integrated)	44.28	46.66	28.20	30.37
Open Ground	12.24	14.33	1.31	2.38
EPBT	Mult-Si	Mono-Si	CIGS L	CIGS S
Slanted Roof (mounted)	60.58	55.32	98.43	98.60
Slanted Roof (integrated)	57.68	52.01	97.62	98.59
Flat Roof (normal)	62.87	57.87	98.79	100.00
Flat Roof (east/west)	46.82	40.52	90.68	92.13
Flat Roof (horizontal)	57.43	51.91	96.08	97.43
Facade (mounted)	15.67	8.01	70.09	70.50
Facade (integrated)	8.01	0.00	64.45	64.84
Open Ground	56.78	52.88	87.82	88.97

Tab. A.11: ΔRem in € /kWp of all considered technology/installation combinations; Source: Authors

	Technology	Slanted Roof (mounted)	Slanted Roof (integrated)	Flat Roof (normal)	Flat Roof (east/west)	Flat Roof (horizontal)	Facade (mounted)	Facade (integrated)	Open Ground
2001	Multi	36.68	16.71	49.68	-37.77	10.67	-116.48	-128.00	1.43
	Mono	34.35	13.92	47.82	-40.04	8.44	-118.01	-129.81	1.41
	CIGS	56.64	40.86	67.33	-22.12	26.46	-106.01	-115.39	2.37
2004	Multi	40.47	18.43	53.81	-40.91	11.56	-95.03	-108.87	33.74
	Mono	37.91	15.36	51.80	-43.37	9.14	-96.87	-111.05	33.32
	CIGS	62.50	45.08	72.92	-23.95	28.66	-82.44	-93.72	55.75
2009	Multi	30.22	13.76	39.82	-30.27	8.55	-95.96	-105.45	23.58
	Mono	28.30	11.47	38.33	-32.09	6.77	-97.23	-106.94	23.29
	CIGS	46.67	33.66	53.96	-17.73	21.21	-87.34	-95.06	38.96
2012	Multi	17.10	7.79	22.33	-16.98	4.80	-54.30	-59.67	13.55
	Mono	16.01	6.49	21.50	-18.00	3.79	-55.01	-60.51	13.38
	CIGS	26.41	19.04	30.26	-9.94	11.89	-49.42	-53.79	22.39
2013	Multi	10.69	4.87	13.24	-10.06	2.84	-33.94	-37.30	7.25
	Mono	10.01	4.06	12.74	-10.67	2.25	-34.39	-37.83	7.16
	CIGS	16.51	11.91	17.94	-5.89	7.05	-30.89	-33.63	11.98

Tab. A.12: Technology specific normalized ΔRem values of all considered installation options; Source: Authors

	Technology	Slanted Roof (mounted)	Slanted Roof (integrated)	Flat Roof (normal)	Flat Roof (east/west)	Flat Roof (horizontal)	Facade (mounted)	Facade (integrated)	Open Ground
2001	Multi	92.68	81.44	100.00	50.78	78.04	6.48	0.00	72.85
	Mono	92.42	80.92	100.00	50.54	77.83	6.64	0.00	73.87
	CIGS	94.15	85.51	100.00	51.05	77.63	5.13	0.00	64.45
	Average	**93.08**	**82.62**	**100.00**	**50.79**	**77.84**	**6.09**	**0.00**	**70.39**
2004	Multi	93.16	81.86	100.00	51.44	78.34	23.69	16.59	89.71
	Mono	92.88	81.32	100.00	51.20	78.13	23.77	16.50	90.53
	CIGS	94.80	86.10	100.00	51.62	77.89	22.41	16.78	91.42
	Average	**93.61**	**83.09**	**100.00**	**51.42**	**78.12**	**23.29**	**16.62**	**90.55**
2009	Multi	93.39	82.06	100.00	51.75	78.48	6.53	0.00	88.82
	Mono	93.10	81.51	100.00	51.52	78.27	6.69	0.00	89.65
	CIGS	95.10	86.37	100.00	51.89	78.02	5.19	0.00	89.94
	Average	**93.86**	**83.32**	**100.00**	**51.72**	**78.26**	**6.14**	**0.00**	**89.47**
2012	Multi	93.62	82.27	100.00	52.06	78.62	6.55	0.00	89.29
	Mono	93.32	81.70	100.00	51.84	78.42	6.71	0.00	90.10
	CIGS	95.41	86.65	100.00	52.17	78.15	5.20	0.00	90.63
	Average	**94.12**	**83.54**	**100.00**	**52.02**	**78.39**	**6.15**	**0.00**	**90.01**
2013	Multi	94.95	83.44	100.00	53.89	79.43	6.64	0.00	88.14
	Mono	94.59	82.82	100.00	53.70	79.25	6.80	0.00	88.95
	CIGS	97.22	88.29	100.00	53.78	78.88	5.30	0.00	88.43
	Average	**95.59**	**84.85**	**100.00**	**53.79**	**79.19**	**6.25**	**0.00**	**88.51**

Tab. A.13: Normalized Δ*Rem* values accounting for technology differences; Source: Authors

	Technology	Slanted Roof (mounted)	Slanted Roof (integrated)	Flat Roof (normal)	Flat Roof (east/west)	Flat Roof (horizontal)	Facade (mounted)	Facade (integrated)	Open Ground
2001	Multi	84.45	74.32	91.05	46.69	71.26	6.76	0.92	66.57
	Mono	83.27	72.91	90.11	45.54	70.13	5.98	0.00	66.57
	CIGS	94.58	86.57	100.00	54.63	79.27	12.07	7.31	67.05
2004	Multi	84.99	74.79	91.16	47.34	71.61	22.30	15.89	81.87
	Mono	83.80	73.37	90.23	46.20	70.49	21.45	14.89	81.68
	CIGS	95.18	87.12	100.00	55.18	79.52	28.12	22.91	92.06
2009	Multi	85.24	75.02	91.21	47.65	71.78	6.82	0.93	81.12
	Mono	84.05	73.59	90.28	46.52	70.67	6.04	0.00	80.94
	CIGS	95.47	87.38	100.00	55.45	79.64	12.19	7.38	90.68
2012	Multi	85.50	75.24	91.26	47.96	71.95	6.84	0.93	81.59
	Mono	84.30	73.81	90.34	46.83	70.84	6.06	0.00	81.40
	CIGS	95.75	87.64	100.00	55.71	79.76	12.22	7.40	91.32
2013	Multi	86.99	76.56	91.57	49.78	72.93	6.96	0.95	80.82
	Mono	85.78	75.10	90.68	48.69	71.86	6.16	0.00	80.66
	CIGS	97.43	89.18	100.00	57.26	80.47	12.44	7.53	89.30

Tab. A.14: MDCA scores applying the AHP method; Source: Authors

AHP scores	Score (Carbon Footprint)	Score (EPBT)	Score (NEA$_{Ep-PCH,LT}$)	Score AHP Method	Normalization
Slanted Roof (mounted)	0.11	0.05	0.06	0.216	100.00
Slanted Roof (integrated)	0.11	0.05	0.05	0.207	95.25
Flat Roof (normal)	0.11	0.05	0.00	0.163	72.00
Flat Roof (east/west)	0.09	0.04	0.04	0.175	78.45
Flat Roof (horizontal)	0.11	0.05	0.05	0.206	94.86
Facade (mounted)	0.03	0.01	0.03	0.067	21.29
Facade (integrated)	0.00	0.00	0.03	0.027	0.00
Open Ground	0.09	0.04	0.02	0.147	63.74

Tab. A.15: MDCA scores applying the weighted sum method; Source: Authors

Weighted sum score	Carbon Footprint	EPBT	NEA$_{Ep-PCH,LT}$	Weighted sum	Normalization
Slanted Roof (mounted)	62.58	7.18	28.28	98.04	100.00
Slanted Roof (integrated)	60.14	6.90	26.42	93.45	94.72
Flat Roof (normal)	64.34	7.38	0.00	71.72	69.69
Flat Roof (east/west)	50.26	5.76	21.37	77.40	76.23
Flat Roof (horizontal)	60.47	6.93	25.61	93.02	94.22
Facade (mounted)	9.99	1.15	12.39	23.53	14.20
Facade (integrated)	0.00	0.00	11.20	11.20	0.00
Open Ground	48.09	5.51	0.42	54.03	49.32

References

Legal References

[2001/77/EC] Directive 2001/77/EC of the European Parliament and of the Council of 27 September 2001 on the promotion of electricity produced from renewable energy sources in the internal electricity market (L283).

[2003/87/EC] Directive 2003/87/EC of the European Parliament and of the Council of 13 October 2003 on establishing a scheme for greenhouse gas emission allowance trading within the Community and amending Council Directive 96/61/EC.

[2009/28/EC] Directive 2009/28/EC of the European Parliament and of the Council of 23 April 2009 on the promotion of the use of energy from renewable sources and amending and subsequently (L140) repealing Directives 2001/77/EC and 2003/30/EC.

[AusglMechV] Ausgleichsmechanismusverordnung vom 17. Juli 2009 (BGBl. I S. 2101), die zuletzt durch Artikel 2 des Gesetzes vom 17. August 2012 (BGBl. I S. 1754) geändert worden ist.

[BT-Drucksache 14/2341] Gesetzentwurf der Fraktionen SPD und BÜNDNIS 90/DIE GRÜNEN: Entwurf eines Gesetzes zur Förderung der Stromerzeugung aus erneuerbaren Energien (Erneuerbare-Energien-Gesetz - EEG) sowie zur Änderung des Mineralölsteuergesetzes. BT-Drucksache 14/2341, 13.12.1999.

[BT-Drucksache 17/8877] Gesetzentwurf der Fraktionen der CDU/CSU und FDP: Entwurf eines Gesetzes zur Änderung des Rechtsrahmens für Strom aus solarer Strahlungsenergie und zu weiteren Änderungen im Recht der erneuerbaren Energien. BT-Drucksache 17/8877, 06.03.2012.

[EEG 2000] Erneuerbare-Energien-Gesetz vom 29. März 2000 (BGBl. I S. 305).

[EEG 2004] Erneuerbare-Energien-Gesetz vom 21. Juli 2004 (BGBl. I S. 1918).

[EEG 2008] Erneuerbare-Energien-Gesetz vom 25. Oktober 2008 (BGBl. I S. 2074).

[EEG 2010-PV] Erstes Gesetz zur Änderung des Erneuerbare-Energien-Gesetzes vom 11. August 2010 (BGBl. I S. 1170).

[EEG 2011] Gesetz zur Umsetzung der Richtlinie 2009/28/EF zur Förderung der Nutzung von Energie aus erneuerbaren Quellen (Europarechtsanpassungsgesetz Erneuerbare Energien - EAG EE) vom 12. April 2011 (BGBl. I S. 619).

[EEG 2012] Erneuerbare-Energien-Gesetz vom 25. Oktober 2008 (BGBl. I S. 2074), das durch Artikel 2 Absatz 69 des Gesetzes vom 22. Dezember 2011 (BGBl. I S. 3044) geändert worden ist.

[EEG 2012-PV] Erneuerbare-Energien-Gesetz vom 25. Oktober 2008 (BGBl. I S. 2074), das durch Artikel 5 des Gesetzes vom 20. Dezember 2012 (BGBl. I S. 2730) geändert worden ist.

[StrEG] Gesetz über die Einspeisung von Strom aus erneuerbaren Energien in das öffentliche Netz (Stromeinspeisungsgesetz) vom 07. Dezember 1990 (BGBl. I S. 2633-2634).

Other References

[AGEE 2012] Arbeitsgemeinschaft Erneuerbare Energien (AGEE). 2012. Zeitreihen zur Entwicklung der erneuerbaren Energien in Deutschland. Excel Sheet. In: http://www.erneuerbare-energien.de/fileadmin/Daten_EE/Bilder_Startseite/Bilder_Datenservice/PDFs__XLS/20130108_EE_Energiedaten_ohne_Formeln_Stand_EEiZIU2011.xls, 06.03.2013.

[Altmaier 2012] Altmaier, P. 2012. Verfahrenvorschlag zur Neuregelung des Erneuerbare-Energien-Gesetzes (EEG). In: www.bmu.de/N49213, 27.02.2013.

[Alsema et al. 2006] Alsema, E. A., de Wild-Scholten, M.,J., Fthenakis, V., M. 2006. Environmental impacts of PV electricity generation - a critical comparison of energy supply options. Proceedings of 21th European Photovoltaic Solar Energy Conference, Dresden, Germany, 4.-8. September 2006.

[Alsema 2012] Alsema, E. 2012. Energy Payback Time and CO_2 Emissions of PV Systems. In: McEvoy, A., Markvart, T., Castaner, L. (Eds.). Practical Handbook of Photovoltaics: Fundamentals and Application. Waltham, Academic Press, pp. 1097-1117.

[Altmaier 2013] Altmaier, P. 2013. Energiewende sichern - Kosten begrenzen: Vorschlag zur Einführung einer Strompreis-Sicherung im EEG. In: www.bmu.de/P2252, 27.02.2013.

[Avril et al. 2012] Avril, S., Mansilla, C., Busson, M., Lemaire, T. 2012. Photovoltaic energy policy: Financial estimation and performance comparison of the public support in five representative countries. Energy Policy, Vol. 51, pp. 244-258.

[BDEW 2013] Bundesverband der Energie- und Wasserwirtschaft (BDEW). 2013. Erneuerbare Energien und das EEG: Zahlen, Fakten, Grafiken (2013). Berlin, BDEW. 80 pp.

[Bechberger 2000] Bechberger, M. 2000. Das Erneuerbare-Energien-Gesetz (EEG): Eine Analyse des Politikformulierungsprozesses. Berlin, Forschungsstelle für Umweltpolitik (FFU), Freie Universität Berlin, FFU-report 00-06. 62 pp.

[Bechberger and Reiche 2004] Bechberger, M., Reiche, D. 2004. Renewable energy policy in Germany: pioneering and exemplary regulations. Energy for Sustainable Development, Vol. VII, No. 1, pp. 47-57.

[Becker and Fischer 2013] Becker, B., Fischer, D. 2013. Promoting renewable electricity generation in emerging economies. Energy Policy, Vol. 56, pp. 446-455.

[BEE 2012] Bundesverband Erneuerbare Energie e.V. (BEE). 2012. BEE-Hintergrund zur EEG-Umlage 2013, Bestandteile, Entwicklung und Höhe: Aktualisierte Fassung nach Veröffentlichung der ÜNB-Prognose vom 15.10.2012. Berlin, BEE. 14 pp.

[Beylot et al. 2012] Beylot, A., Payet, J., Puech, C., Adra, N., Jacquin, P., Blanc, I., Beloin-Saint-Pierre, D. Environmental impacts of large-scale grid-connected ground-mounted PV installations. Renewable Energy, Forthcoming, pp. 1-5.

[BMU 2007] Bundesministerium für Umwelt, Naturschutz und Reaktorsicherheit (BMU). 2007. Erfahrungsbericht 2007 zum Erneuerbare-Energien-Gesetz (EEG-Erfahrungsbericht). Berlin, BMU. 186 pp.

[BMU 2011] Bundesministerium für Umwelt, Naturschutz und Reaktorsicherheit (BMU). 2011. Das Energiekonzept der Bundesregierung 2010 und die Energiewende 2011. Berlin, BMWi. 32 pp.

[BMU 2012a] Bundesministerium für Umwelt, Naturschutz und Reaktorsicherheit (BMU). 2012a. Entwicklung der erneuerbaren Energien in Deutschland im Jahr 2011. In: http://www.erneuerbare-energien.de/fileadmin/ee-import/files/pdfs/allgemein/application/pdf/ee_in_deutschland_graf_tab.pdf, 27.02.2013.

[BMU 2012b] Bundesministerium für Umwelt, Naturschutz und Reaktorsicherheit (BMU). 2012a. Photovoltaik: Einigung im Vermittlungsausschuss. Press Release, Nr. 096/12, Berlin, 28.06.2012.

[BMU 2012c] Bundesministerium für Umwelt, Naturschutz und Reaktorsicherheit (BMU). 2012c. Erneuerbare Energien in Zahlen: Nationale und Internationale Entwicklung. Berlin, BMU. 134 pp.

[BMWi and BMU 2012a] Bundesministerium für Wirtschaft und Technologie (BMWi) and Bundesministerium für Umwelt, Naturschutz und Reaktorsicherheit (BMU). 2012a. EU-Effizienzrichtlinie und Erneuerbare-Energien-Gesetz: Ergebnispapier. Berlin, BMWi and BMU. 11 pp.

[BMWi and BMU 2012b] Bundesministerium für Wirtschaft und Technologie (BMWi) and Bundesministerium für Umwelt, Naturschutz und Reaktorsicherheit (BMU). 2012b. Erster Monitoring-Bericht: "Energie der Zukunft". Berlin, BMWi and BMU. 125 pp.

[Boute 2012] Boute, A. 2012. Promoting renewable energy through capacity markets: An analysis of the Russian support scheme. Energy Policy, Vol. 46, pp. 68-77.

[Bruckner et al. 2010] Bruckner, T., Edenhofer, O., Held, H., Haller, M., Lüken, M., Bauer, N., Nakicenovic, N. 2010. Robust options for decarbonisation. In: Schellnhuber, H., J., Molina, M., Stern, N., Huber, V., Kadner, S. (Eds.). Global Sustainability – A Nobel Cause. Cambridge and New York, Cambridge University Press. pp. 189-204.

[BSW-Solar 2012] Bundesverband Solarwirtschaft (BSW-Solar). 2012. Mehr als 11.000 Menschen protestierten in Berlin gegen "Solar-Ausstieg". Press Release, Berlin, 05.03.2012.

[Bundesnetzagentur 2013] Bundesnetzagentur. 2013. Photovoltaikanlagen: Datenmeldungen sowie EEG-Vergütungssätze. In: http://www.bundesnetzagentur.de/cln_1911/DE/Sachgebiete/ElektrizitaetundGas/Unternehmen_Institutionen/ErneuerbareEnergien/Photovoltaik/DatenMeldgn_EEG-VergSaetze/DatenMeldgn_EEG-VergSaetze_node.html, 06.03.2013.

[Bundesregierung 2002a] Bundesregierung. 2002a. Bericht über den Stand der Markteinführung und der Kostenentwicklung von Anlagen zur Erzeugung von Strom aus erneuerbaren Energien (Erfahrungsbericht zum EEG). Berlin, Bundesregierung. 42 pp.

[Bundesregierung 2002b] Bundesregierung. 2002b. Perspektiven für Deutschland: Unsere Strategie für eine nachhaltige Entwicklung. Berlin, Bundesregierung. 40 pp.

[Bundesregierung 2010] Bundesregierung. 2010. Energiekonzept für eine umweltschonende, zuverlässige und bezahlbare Energieversorgung. Berlin, Bundesregierung. 328 pp.

[Bundesregierung 2011] Bundesregierung. 2011. Erfahrungsbericht 2011 zum Erneuerbare-Energien-Gesetz (EEG-Erfahrungsbericht). Berlin, Bundesregierung. 23 pp.

[Bundesregierung 2012] Bundesregierung. 2012. Das neue Marktintegrationsmodell für Strom aus solarer Strahlungsenergie im Erneuerbare-Energien-Gesetz (EEG). In: `http://www.erneuerbare-energien.de/fileadmin/ee-import/files/pdfs/allgemein/application/pdf/marktintegrationsmodell_bf.pdf`, 06.03.2013.

[Butler and Neuhoff 2008] Butler, L., Neuhoff, K. 2008. Comparison of feed-in tariff, quota and auction mechanisms to support wind power development. Renewable Energy, Vol. 33, pp. 1854-1867.

[Cardoso Marques and Fuinhas 2012] Cardoso Marques, A., Fuinhas, J., A. Are public policies towards renewables successful? Evidence from European countries. Renewable Energy, Vol. 44, pp. 109-118.

[Couture and Gagnon 2010] Couture, T., Gagnon, Y. An analysis of feed-in tariff remuneration models: Implications for renewable energy investment. Energy Policy, Vol. 38, pp. 955-965.

[CRE 2013a] Commission de Régulation de L'Énergie (CRE). 2013. Cahier des charges de l'appel d'offers portant sur la réalisation et l'exploitation d'installations photovoltaiques sur bâtiment de puissance crête comprise entre 100 et 250 kW. Paris, CRE. 28 p. To be found in: `http://www.cre.fr/documents/appels-d-offres/`, 06.04.2013.

[CRE 2013b] Commission de Régulation de L'Énergie (CRE). 2013. Cahier des charges de l'appel d'offres portant sur la réalisation et l'exploitation d'installations de production d'électricité à partir de l'énergie solaire d'une puissance supérieure à 250 kWc. Paris, CRE. 61 p. To be found in: `http://www.cre.fr/documents/appels-d-offres/`, 06.04.2013.

[de Wild-Scholten and Alsema 2005] de Wild-Scholten, M., J., Alsema, E., A. 2007. Environmental Life Cycle Inventory of Crystalline Silicon Photovoltaic System Production. Petten, Energy research Center of the Netherlands and Utrecht, Copernicus Institute for Sustainable Development and Innovation. Excel-Sheet to be found in: `http://www.ecn.nl/publicaties/ECN-E--07-026`, 27.04.2013.

[de Wild-Scholten 2011] de Wild-Scholten, M., J. 2011. Environmental profile of PV mass production: globalization. Presentation at the 26th European Photovoltaic Solar Energy Conference, Hamburg, 8. September 2011.

[del Río 2012] del Río, P. 2012. The dynamic efficiency of feed-in tariffs: The impact of different design elements. Energy Policy, Vol. 41, pp. 139-151.

[Desideri et al., 2012] Desideri, U., Proietti, S., Zepparelli, F., Sdringola, P. Bini, S. Life Cycle Assessment of a ground-mounted 1778 kWp photovoltaic plant and comparison with traditional energy production systems. Applied Energy, No. 97, pp. 930-943.

[ECJRC n.d.] European Commission Joint Reseach Centre (ECJRC). n.d. Photovoltaic Geographical Information System - Interactive Maps. In: http://re.jrc.ec.europa.eu/pvgis/apps4/pvest.php, 13.03.2013.

[ECJRC 2010a] European Commission Joint Research Centre (ECJRC). 2010a. International Reference Life Cycle Data System (ILCD) Handbook: General guide for Life Cycle Assessment: Detailed guidance. Luxembourg, Publications Office of the European Union, first edition, EUR 24708 EN - 2010. 394 pp.

[ECJRC 2010b] European Commission Joint Research Centre (ECJRC). 2010b. International Reference Life Cycle Data System (ILCD) Handbook: Framework and requirements for LCIA models and indicators. Luxembourg, Publications Office of the European Union, first edition, EUR 247586 EN - 2010. 103 pp.

[EEA 1999] European Energy Agency (EEA). 1999. Environmental indicators: Typology and overview. Copenhagen, EEA, Technical report No. 25, 1999. 19 pp.

[EEA 2003] European Energy Agency (EEA). 2003. Environmental indicators: Typology and overview. Copenhagen, EEA, Update on Technical report No. 25, 1999. 20 pp.

[EEA 2005] European Energy Agency (EEA). 2005. EEA core set of indicators: Guide. Copenhagen, EEA, Technical report No. 1, 2005. 38

pp.

[EEA 2012] European Energy Agency (EEA). 2012. Climate change, impacts and vulnerability in Europe 2012: An indicator-based report. Copenhagen, EEA Report No. 12, 2012. 300 pp.

[EEG-Explanation 2004] EEG-Explanation. 2004. Konsolidierte Fassung der Begründung zu dem Gesetz für den Vorrang Erneuerbarer Energien (Erneuerbare-Energien-Gesetz – EEG) vom 21. Juli 2004 (BGBl. 2004 I S. 1918). In: http://www.clearingstelle-eeg.de/files/private/active/0/EEG_2004_begruendung_konsolidiert.pdf, 04.03.2013.

[EEG-Explanation 2009] EEG-Explanation. 2009. Konsolidierte Fassung der Begründung zu dem Gesetz für den Vorrang Erneuerbarer Energien (Erneuerbare-Energien-Gesetz – EEG) vom 25. Oktober 2008 (BGBl. I S. 2074). In: http://www.clearingstelle-eeg.de/files/A10-EEG_2009_konsolidierte_Begr_0.pdf, 04.03.2013.

[EEG-Explanation 2012] EEG-Explanation. 2012. Konsolidierte Fassung der Begründung zu dem Gesetz für den Vorrang Erneuerbarer Energien (Erneuerbare-Energien-Gesetz – EEG) vom 25. Oktober 2008 (BGBl. I S. 2074), das durch Artikel 2 Absatz 69 des Gesetzes vom 22. Dezember 2011 (BGBl. I S. 3044) geändert worden ist. In: https://www.mitnetz-strom.de/data/Begruendung_EEG-2012_Stand_06.06.2011.pdf, 04.03.2013.

[EFI 2013] Expertenkommission Forschung und Innovation (EFI). 2013. Gutachten zu Forschung, Innovation und technologischer Leistungsfähigkeit Deutschlands. In: http://www.e-fi.de/fileadmin/Gutachten/EFI_2013_Gutachten_deu.pdf, 27.02.2013.

EN ISO 14040:2006. Environmental management - Life cycle assessment - Principles and framework. DIN Deutsches Institut für Normung e.V. Beuth Verlag, Berlin, 44 pp.

EN ISO 14044:2006. Environmental management - Life cycle assessment - Requirements and guidelines. DIN Deutsches Institut für Normung e.V. Beuth Verlag, Berlin, 84 pp.

[European Commission 2008] European Commission. 2008. 20 20 by 2020 Europe's climate change opportunity. Brussels, European Commission, Communication from the Commission to teh European Parliament, the Council, the European Economic and Social Committee of the Regions, COM (2008) 30.

[EPIA 2011] European Photovoltaic Industry Association (EPIA). 2011. Solar Generation 6: Solar photovoltaic electricity empowering the world. In: http://www.greenpeace.org/international/ Global/international/publications/climate/2011/ Final%20SolarGeneration%20VI%20full%20report% 20lr.pdf, 14.03.2013.

[Ferraza 2012] Ferraza, F. 2012. Crystalline Silicon: Manufacture and Properties. In: McEvoy, A., Markvart, T., Castaner, L. (Eds.): Practical Handbook of Photovoltaics: Fundamentals and Application. Waltham, Academic Press, pp. 79-97.

[Finon and Perez 2007] Finon, D., Perez, Y. 2007. The social efficiency of instruments of promotion of renewable energies: A transaction-cost perspective. Ecological Economics, Vol. 62, pp. 77-92.

[Fouquet and Johansson 2008] Fouquet, D., Johansson, T., B. 2008. European renewable energy policy at crossroads - Focus on electricity support mechanisms. Energy Policy, Vol. 36, pp. 4079-4092.

[Frischknecht et al. 2007] Frischknecht, R., Jungbluth, N., Althaus, H., J., Doka, G., Dones, R., Hirschier, R., Hellweg, S., Humbert, S., Margni, M., Nemecek, T., Spielmann, M. 2007. Overview and Methodology. Dübendorf, Swiss Centre for Life Cycle Inventories, Eco-Invent Report No. 1. 68 pp.

[Frondel et al. 2008] Frondel, M., Ritter, N., Schmidt, Chr. M. 2008. Germany's solar cell promotion: Dark clouds on the horizon. Energy Policy, Vol. 36, pp. 4198-4204.

[Frondel et al. 2010] Frondel, M., Ritter, N., Schmidt, C., M., Vance, C. 2010. Economic impacts from the promotion of renewable energy technologies: The German experience. Energy Policy, Vol. 38, pp. 4048-4056.

[Fthenakis and Alsema 2005] Fthenakis V., Alsema E. 2005. Photovoltaics energy payback times, greenhouse gas emissions and external costs: 2004-2005 status. Progress in Photovoltaics: Research and Applications, Vol. 14, pp. 275-278.

[Fthenakis and Kim 2011] Fthenakis, V., Kim, H. C. 2011. Photovoltaics: Life-cycle analyses. Solar Energy, Vol. 85, pp. 1609-1628.

[Gawel et al. 2012a] Gawel, E., Korte, K., Lehmann, P., Strunz, S. 2012. The German Energy Transition – Is It Really Scandalous? False Alarm! Neither Command Economy Nor "Cost Tsunami" Are Imminent. GAiA, Vol. 21, Issue 4, pp. 278-283.

[Gawel et al. 2012b] Gawel, E., Lehmann, P., Strunz, S., Korte, K. 2012. Kosten der Energiewende - Fakten und Mythen. Energiewirtschaftliche Tagesfragen, Vol. 62, Issue 11, pp. 39-44.

[Gawel and Purkus 2013] Gawel, E., Purkus, A. 2013. Promoting the Market and System Integration of Renewable Energies through Premium Schemes – A Case Study of the German Market Premium. Energy Policy, Vol. 61, pp. 599-609.

[Gawel et al. 2013a] Gawel, E., Strunz, S., Lehmann, P. 2013a. Germany's energy transition under attack. Is there an inscrutable German *Sonderweg*? Nature and Culture, Vol. 8, Issue 2, pp. 121-133.

[Gawel et al. 2013b] Gawel, E., Strunz, S., Lehmann, P. 2013b. A public choice view on the climate and energy policy mix in the EU – How do Emissions Trading Scheme and support for renewable energies interact?. Energy Policy, forthcoming (DOI: 10.1016/j.enpol.2013.09.008).

[Goedkoop et al. 2009] Goedkoop, M., Heijungs, R., Huijbregts, M., Schryver, A., D., Struijs, J., Zelm, R. 2009. ReCiPe 2008: A life cycle impact assessment method which comprises harmonised category indicators at the midpoint and the endpoint level.

In: http://www.pre-sustainability.com/download/ misc/ReCiPe_main_report_final_27-02-2009_web.pdf, 09.03.2013.

[Griffin 2013] Griffin, R. 2013. Auction designs for allocating wind energy leases on the U.S. outer continental shelf. Energy Policy, Vol. 56, pp. 603-611.

[Haas et al. 2011] Haas, R., Panzer, C., Resch, G., Ragwitz, M., Reece, G., Held, A. 2011. A historical review of promotion strategies for electricity from renewable energy sources in EU countries. Renewable and Sustainable Energy Reviews, Vol. 15, pp. 1003-1034.

[Halasah et al. 2013] Halasah, S., A., Pearlmutter, D., Feuermann, D. 2013. Field installation versus local integration of photovoltaic system and their effect on energy evaluation metrics. Energy Policy, Vol. 52, pp. 462-471.

[Hegedus and Luque 2003] Hegedus, S., Luque, A. 2003. Status, Trends, Challenges and the Bright Future of Solar Electricity from Photovoltaics. In: Luque, A. and Hegedus, S. (Eds.): Handbook of Photovoltaic Science and Engineering. London, Wiley, pp. 1-43.

[Hischier et al. 2010] Hischier R., Weidema, B., Althaus, H., J., Bauer, C., Doka, G., Dones, R., Frischknecht, R., Hellweg, S., Humbert, S., Jungbluth, N., Köllner, T., Loernick, Y., Margni, M., Nemecek, T. 2010. Implementation of Life Cycle Impact Assessment Methods. Dübendorf, Swiss Centre for Life Cycle Inventories, ecoinvent report No. 3, v2.2.

[Hoppmann et al. 2013] Hoppmann, J., Peters, M., Schneider, M., Hoffmann, V., H. 2013. The two faces of market support - How deployment policies affect technological exploration and exploitation in the solar photovoltaic industry. Research Policy, Vol. 42, pp. 989-1003.

[IEA 2008] International Energy Agency (IEA). 2008. Deploying Renewables: Principles for Effective Policies. Paris, IEA. 250 pp.

[IEA 2010] International Energy Agency (IEA). 2010. Energy Technology Perspectives: Scenarios & Strategies to 2050. Paris, IEA. 706 pp.

[IEAPVPS 2011] International Energy Angency Photovoltaic Power System Programme (IEAPVPS). 2011. Methodology Guidelines on Life Cycle Assessment of Photovoltaic Electricity. In: http://www.iea-pvps.org/fileadmin/dam/public/ report/technical/rep12_11.pdf, 18.02.2013.

[IPCC 2007] Intergovernmental Panel on Climate Change (IPCC). 2007. Climate Change 2007: Synthesis Report. Geneva, IPCC. 104 pp.

[IPCC 2012] Intergovernmental Panel on Climate Change (IPCC). 2012. Renewable Energy Sources and Climate Change Mitigation: Special Report of the Intergovernmental Panel on Climate Change. Cambridge et al., Cambridge University Press. 1076 pp.

[Ito et al. 2003] Ito M., Kato, K., Sugihara, H., Kichimi, T., Song, J., Kurokawa, K. A preliminary study on potential for very large-scale photovoltaic power generation (VLS-PV) system in the Gobi desert from economic and environmental viewpoints. Solar Energy Material & Solar Cells Vol. 75, pp. 505-517.

[Ito et al. 2010] Ito, M., Komoto, K., Kurokawa, K. 2010. Life-cycle analyses of very-large scale PV systems using six types of PV modules. Current Applied Physics, Vol. 10, pp. 5271-5273.

[Jäger-Waldau 2012] Jäger-Waldau, A. 2012. PV Status Report 2012: Part 1. Luxembourg, Publications Office of the European Union, Report EUR 25749 EN. 45 pp.

[Jenner et al. 2013] Jenner, S., Groba, F., Indvik, J. Assessing the strength and effectiveness of renewable electricity feed-in tariffs in European Union countries. Energy Policy, Vol. 52, pp. 385-401.

[Jungbluth et al. 2005] Jungbluth, N., Bauer, C., Dones, R., Frischknecht, R. 2005. Life Cycle Assessment for Emerging Technologies: Case Studies for Photovoltaic and Wind Power. International Journal of Life Cycle Assessment, Vol. 10, pp. 24-34.

[Jungbluth et al. 2010] Jungbluth, N., Stucki, M., Frischknecht, R., Büsser, S. 2010. Photovoltaics. In: Dones, R. et al. (Eds.). Sachbilanzen von Energiesystemen: Grundlagen für den Ökologischen Vergleich von Energiesystemen und den Einbezug von Energiesystemen in Ökobilanzen für die Schweiz. EcoInvent report No. 6-XII. Uster, ESU-services Ltd. 228 pp.

[Kato et al. 1998] Kato, K., Murata, A., Sakuta, K. Energy pay-back time and life-cycle CO_2 emission of residential PV power system with silicon PV module. Progress in Photovoltaics Research and Applications, Vol. 6, pp. 105–115.

[Kersten et al. 2011] Kersten, F., Doll, R., Kux, A., Huljic, D., M., Görig, M., A., Breyer, C., Müller, J., W., Wawer, P. PV learning curves: past and future drivers of cost reduction. Proceedings of 26th European Photovoltaic Solar Energy Conference, Hamburg, Germany, 5.-9. September 2011.

[KfW n.d.] Kreditanstalt für Wiederaufbau. N.d. Erneuerbare Energien - Standard - Photovoltaik. In: https://www.kfw.de/inlandsfoerderung/ Privatpersonen/Neubau/Finanzierungsangebote/ Erneuerbare-Energien-Standard-(274)/#2, 08.04.2013.

[Kim and Lee 2012] Kim, K., K., Lee, C., G. 2012. Evaluation and optimization of feed-in tariffs. Energy Policy, Vol. 49, pp. 192-203.

[Klein et al. 2008] Klein, A., Pfluger, B., Held, A., Ragwitz, M., Resch, G., Faber, T. 2008. Evaluation of different feed-in tariff design options - Best practice paper for the International Feed-In Cooperation, 2nd Edition. Karlsruhe, Fraunhofer Institut für System- und Innovationsforschung, Energy Economics Group. 87 pp.

[Klessmann et al. 2008] Klessmann, C., Nabe, C., Burges, K. 2008. Pros and cons of exposing renewables to electricity market risks - A comparison of the market integration approaches in Germany, Spain, and the UK. Energy Policy, Vol. 36, pp. 3646-3661.

[Langniß et al. 2009] Langniß, O., Diekmann, J., Lehr, U. 2009. Advanced mechanisms for the promotion of renewable energy - Models for the future evolution of the German Renewable Energy Act. Energy Policy, Vol. 37, pp. 1289-1297.

[Laleman et al. 2011] Laleman, R., Albrecht, J., Dewulf, J. Life Cycle Analysis to estimate the environmental impact of residential photovoltaic systems in regions with a low solar irradiation. Renewable and Sustainable Energy Reviews, Vol. 15, pp. 267-281.

[Lehmann 2012] Lehmann, P. 2012. Justifying a policy mix for pollution control: A review of economic literature. Journal of Economic Surveys, Vol. 26, No. 1, pp. 71-97.

[Lehmann et al. 2012] Lehmann, P, Creutzig, F., Ehlers, M.-H., Friedrichsen, N., Heuson, C., Hirth L., Pietzcker, R. 2012. Carbon Lock-Out: Advancing Renewable Energy Policy in Europe. Energies, Vol. 5, Issue 2, pp. 323-354.

[Lehmann and Gawel 2013] Lehmann, P., Gawel, E. 2013. Why should support schemes for renewable electricity complement the EU emissions trading scheme? Energy Policy, Vol. 52, pp. 597-607.

[Lipp 2007] Lipp, J. 2007. Lessons for effective renewable electricity policy from Denmark, Germany and the United Kingdom. Energy Policy, Vol. 35, pp. 5481-5495.

[Lu and Yang 2010] Lu, L., Yang., H., X. 2010. Environmental payback time analysis of a roof-mounted building-integrated photovoltaic (BIPV) system in Hong Kong. Applied Energy, Vol. 87, pp. 3625-3631.

[Mateo 2012] Mateo, J., R., S., C. 2012. Multi Criteria Analysis in the Renewable Energy Industry. Dordrecht, Springer. 107 pp.

[MEA 2005] Millennium Ecosystem Assessment (MEA). 2005. Ecosystems and Human Well-being: Current State and Trends, Volume 1. Washington et al., Island Press. 23 pp.

[Mendonca et al., 2010] Mendonca, M., Jacobs, D., Sovacool, B. 2010. Powering the Green Economy: The feed-in tariff handbook. London, Earthscan. 208 pp.

[Mennel 2012] Mennel, T. 2012. Das Erneuerbare-Energien-Gesetz - Erfolgsgeschichte oder Kostenfalle? Wirtschaftsdienst Vol. 92, No. 13, pp. 17-22.

[McCool and Stankey 2004] McCool, S., F., Stankey, G., H. 2004. Indicators of Sustainability: Challenges and Opportunities at the Interface of Science and Policy. Environmental Management, Vol. 33, No. 3, pp. 294-305.

[McDonald and Pearce 2010] McDonald, N., C., Pearce, J., M. 2010. Producer responsibility and recycling solar photovoltaic modules. Energy Policy, Vol. 38, pp. 7041-7041.

[Mason et al., 2005] Mason, J., E., Fthenakis, V., M., Hansen, T., Kim, H., C. 2005. Energy Payback and Life-cycle CO_2 Emissions of the BOS in an Optimized 3.5 MW PV installation. Progress in Photovoltaics: Research and Applications, No. 14, pp. 179-190.

[Nelskamp 2012] Nelskamp. 2012. Prospekt Doppelmuldenfalz-Ziegel D 13, Stand 10/2012. In: http://www.nelskamp.de/fileadmin/downloads/de/d13/d13_prospekt_de.pdf, 03.12.2012.

[Niemeijer and de Groot 2008] Niemeijer, D., de Groot, R., S. 2008. A conceptual framework for selecting environmental indicator sets. Ecological Indicators, Vol. 8, pp. 14-25.

[Nishimura et al. 2010] Nishimura, A., Hayashi, Y., Tanaka, K., Hirota, M., Kato, S., Ito, M., Araki, K., Hud, E., J. 2010. Life cycle assessment and evaluation of energy payback time on high-concentration photovoltaic power generation system. Applied Energy, Vol. 87, pp. 2797-2807.

[NRC 2000] United States National Research Council (NRC). 2000. Ecological Indicators for the Nation. Washington D. C., National Acadamy Press. 180 pp.

[NREL n.d.] United States National Renewable Energy Laboratory (NREL). n.d. Solar Resource Models and Tools: PVWatts: A Performance Calculator for Grid-Connected PV Systems. In: http://www.nrel.gov/rredc/models_tools.html, 13.03.2013.

[Pacca et al. 2007] Pacca, S., Sivaraman, D., Keoleian, G., A. 2007. Parameters affecting the life cycle performance of PV technologies and systems. Energy Policy, Vol. 35, pp. 3316-3326.

[Peng et al. 2013] Peng, J., Lu, L., Yang, H. Review on life cycle assessment of energy payback and greenhouse gas emission of solar photovoltaic systems. Renewable and Sustainable Energy Reviews, Vol. 19, pp. 255-274.

[Philibert 2011] Philibert, C. 2011. Interactions of Policies for Renewable Energy and Climate. Paris, IEA Working Paper. 22 pp.

[Preiser 2003] Preiser, K. 2003. Photovoltaic Systems. In: Luque, A. and Hegedus, S. (Eds.): Handbook of Photovoltaic Science and Engineering. London, Wiley, pp. 753-798.

[Hanwha Q CELLS n.d.] Hanwha Q CELLS GmbH. n. d. Solarzellen, Solarmodule und Solaranlagen. In http://www.q-cells.com/solarzellen_solarmodule_und_solaranlagen.html, 16.11.2012.

[SENSE, 2008] Sustainability Evaluation of Solar Energy Systems (SENSE). 2008. LCA Analysis. Revised Version 06/2008. In: http://www.sense-eu.net/fileadmin/user_upload/intern/documents/Results_and_Downloads/SENSE_LCA_results.pdf, 29.11.2012.

[Ragwitz et al. 2007] Ragwitz, Held, A., Resch, G., Faber, T., Haas, R., Huber, C., Coenraad, R., Voogt, M., Reece, G., Morthorst, P., E., Grenaa Jensen, S., Konstantinaviciute, I., Heyder, B. 2007. Assessment and optimization of renewable energy support schemes in European electricity market. Karlsruhe, European Commission, Intelligent Energy Executive Agency, OPTRES Final Report. 226 pp.

[Rametsteiner et al. 2011] Rametsteiner, E., Pülzl, H., Alkan-Olsson, J., Frederiksen, P. 2011. Sustainability indicator development - Science or political negotiation? Ecological indicators, Vol. 11, pp. 61-70.

[Raugei et al. 2007] Raugei, M., Bargigli, S., Ulgiati, S. 2006. Life cycle assessment and energy pay-back time of advanced photovoltaic modules: CdTe and CIS compared to poly-Si. Energy, Vol. 32, pp. 1310-1318.

[Reichmuth 2011] Reichmuth, M. 2011. Vorbereitung und Begleitung der Erstellung des Erfahrungsberichtes 2011 gemäß § 65 EEG:

Vorhaben IIc Solare Strahlungsenergie. Berlin, BMU. 397 pp.

[Reijenga 2003] Reijenga, T., H. 2003. PV in Architecture. In: Luque, A. and Hegedus, S. (Eds.): Handbook of Photovoltaic Science and Engineering. London, Wiley, pp. 1005-1042.

[REN21 2012] Renewable Energy Policy Network for the 21th Century (REN21). Renewables 2012: Global Status Report. Paris, REN21. 171 pp.

[Roßegger 2008] Roßegger, U. 2008. Die Förderung erneuerbarer Energien in der Europäischen Union: Ein Vergleich der Förderinstrumente anhand der Fallbeispiele Deutschland und Großbritannien. Nerderstedt, GRIN Verlag. 136 pp.

[Schmidt 2012] Schmidt, M. 2012. Tariff Setting for the promotion of Renewable Energy - General Approach. Supportive Policy Framework for Renewable Energy - Experiences for the design of Feed in-Tariff Systems, Tokyo, 06.03.2012. Stuttgart, Zentrum für Sonnenenergie - und Wasserstoff-Forschung Baden-Württemberg (ZSW). 17 pp.

[Sensfuß 2011a] Sensfuß, F. 2011a. Analysen zum Merit-Order Effekt erneuerbarer Energien: Update für das Jahr 2010. Karlsruhe, Fraunhofer Institut für System- und Innovationsforschung. 20 pp.

[Sensfuß 2011b] Sensfuß, F. 2011b. Vorbereitung und Begleitung der Erstellung des Erfahrungsberichtes 2011 gemäß §65 EEG: Vorhaben IV Instrumente und rechtliche Weiterentwicklung im EEG. Berlin, BMU. 523 pp.

[Sherwani et al. 2010] Sherwani, A., F., Usmani, J., A., Varun. Life cycle assessment of solar PV based electricity generation system: A review. Renewable and Sustainable Energy Reviews, Vol. 14, pp. 540-544.

[Solibro n.d.] Solibro GmbH. N. d. Solibro Solarpower. In http://solibro-solarpower.com/, 16.11.2012.

[Sorrell and Sijm 2003] Sorrell, S., Sijm, J. 2003. Carbon Trading in the Policy Mix. Oxford Review of Economic Policy, Vol. 19, No. 3, pp. 420-437.

[SPD/Grüne 1998] Sozialdemokratische Partei Deutschlands und Bündnis 90/Die Grünen. 1998. Aufbruch und Erneuerung - Deutschlands Weg ins 21. Jahrhundert: Koalitionsvereinbarung zwischen der Sozialdemokratischen Partei Deutschlands und Bündnis 90/Die Grünen. Bonn, SPD and BÜNDNIS 90/DIE GRÜNEN. 51 pp.

[Staiß et al. 2007] Staiß, F., Schmidt, M., Musiol, F. Forschungsvorhaben im Auftrag des Bundesministeriums für Umwelt, Naturschutz und Reaktorsicherheit: Vorbereitung und Begleitung der Erstellung des Erfahrungsberichtes 2007 gemäß §20 EEG. Berlin, BMU. 510 p.

[Sumper et al. 2011] Sumper, A., Robledo-García, M., Villafafila-Robles, R., Bargas-Jan, J., Andrs-Peir, J. 2011. Life-cycle assessment of a photovoltaic system Catalonia (Spain). Renewable and Sustainable Energy Reviews, Vol. 15., pp. 3888-3896.

[Swarr et al. 2011] Swarr, T., E., Hunkeler, D., Klopffer, W., Pesonen H., L., Ciroth, A., Brent, A., C., Pagan, R. 2011. Environmental Life Cycle Costing: A Code of Practice. Pensacola, Society of Environmental Chemistry and Toxicology (SETAC). 98 pp.

[Szlufcik et al. 2012] Szlufcik, J., Sivoththaman, S., Nijs, J., F., Mertens, R., P., van Overstraeten, R. 2012. Low-Cost Industrial Technologies for Crystalline Silicon Solar Cells. In: McEvoy, A., Markvart, T., Castaner, L. (Eds.): Practical Handbook of Photovoltaics: Fundamentals and Application. Waltham, Academic Press, pp. 129-159.

[Tobías et al. 2003] Tobías, I., Canzio, C., Alonso, J. 2003. Crystalline Silicon Solar Cells and Modules. In: Luque, A. and Hegedus, S. (Eds.): Handbook of Photovoltaic Science and Engineering. London, Wiley, pp. 255-306.

[Töpfer 2012] Töpfer, C. 2012. Is the German photovoltaic subsidization scheme goal oriented? An environmental profile of eight different PV installation using different module technologies. Unpublished Paper for the *Integration Module of the Joint International Master of Sustainable Development* at the University of Leipzig, Leipzig. 106 pp.

[TSO 2012] Transmission System Operators (TSO). 2012. EEG-Mittelfristprognose: Entwicklungen 2013 bis 2017 (Trend-Szenario). In: http://www.eeg-kwk.net/de/file/Zusammenfassung_Mifri_Einspeisung_2013_-_2017.pdf, 06.03.2013.

[Turney and Fthenakis 2011] Turney, D., Fthenakis, V. 2011. Environmental impacts from the installation and operation of large-scale solarpower plants. Renewable and Sustainable Energy Reviews, Vol. 15, pp. 3161-3270.

[UBA 2009] Umweltbundesamt (UBA). 2009. Emissionsbilanz erneuerbarer Energieträger: Durch Einsatz erneuerbarer Energien vermiedene Emissionen im Jahr 2007. In: http://www.umweltdaten.de/publikationen/fpdf-l/3761.pdf, 27.04.2013.

[UBA 2011] Umweltbundesamt (UBA). 2011. Stromerzeugung aus erneuerbaren Energien: klimafreundlich und ökonomisch sinnvoll. Dessau-Roßlau, UBA. 19 pp.

[UNEP 2009] United Nations Environmental Program (UNEP). 2009. Guidelines for Social Life Cycle Assessment of Products. Nairobi, UNEP. 101 pp.

[UNEP 2010] United Nations Environmental Programme (UNEP). 2010. Assessing the Environmental Impacts of Consumption and Production: Priority Products and Materials. Nairobi, UNEP. 108 pp.

[UNFCCC 2011] United Nations Framework Conference on Climate Change (UNFCCC). 2011. Report of the Conference of the Parties on its sixteenth session, held in Cancun from 29 November to 10 December 2010, Part Two: Action taken by the Conference of the Parties at its sixteenth session, Contents: Decisions adopted by the Conference of the Parties. UNFCC, FCCC/CP/2010/7/Add.1. 31 pp.

[VDI 2067:2012] Economic efficiency of building installations - Fundamentals and economic calculation. VDI Verein Deutscher Ingenieure. Beuth Verlag, Berlin, 44 pp.

[Verbruggen and Lauber 2012] Verbruggen, A., Lauber, V. 2012. Assessing the performance of renewable electricity support instruments. Energy Policy, Vol. 45, pp. 635-644.

[WBCSD 2000] World Business Council for Sustainable Development (WBCSD). 2000. Eco-efficiency: Creating more value with less impact. Geneva, WBCSD. 32 pp.

[WBGU 2009] Wissenschaftlicher Beirat der Bundesregierung Globale Umweltveränderungen (WBGU). 2009. Kassensturz für den Weltklimavertrag: Der Budgetansatz. Berlin, WBGU, Sondergutachten. 58 pp.

[Wissing 2012] Wissing, L. 2012. National Survey Report of PV Power Applications in Germany 2011. International Energy Agency Co-Operative Programme on Photovoltaic Power Systems: Task 1 Exchange and dissemination of information on PV power systems. In: http://www.iea-pvps.org/index.php?id=93&eID=dam_frontend_push&docID=1234, 08.04.2013.

[Zuser and Rechberger 2011] Zuser, A., Rechberger, H. 2011. Considerations of resource availability in technology development strategies: The case study of photovoltaics. Resources, Conservation and Recycling, Vol. 56, pp. 56-65.